Preface

D1363968

"Mathematics is the language with which God has written the universe"

\- Galileo Galilei

Especially when learned during one's formative years, mathematics can serve as a foundation for success in many areas. Learning mathematics allows us to decipher the world around us through pattern recognition and logical reasoning. In particular, the subject of geometry is a powerful conduit toward the teaching of critical thinking and problem-solving skills.

This book is aimed toward middle- and high-school students who enjoy participating in math contests and want to master the topic of geometry. In this approach, I have tried to make learning feel effortless and exciting by providing concise knowledge sections coupled with thought-provoking practice problems.

This book has two main sections: a theory section with 6 units and a practice section with 225 problems. Each problem has several detailed solutions that contain diagrams to aid in the students' visualization process. The book also provides a hint box below each question that gives guidance toward the recommended method of solving the problem, in case of difficulty.

I wrote this book based on my personal experience from participating in math competitions, and the topics I've noticed students commonly struggle with. I hope that by the end of the book, readers will feel more confident in their critical thinking abilities and further their appreciation for geometry.

Acknowledgements

I want to start by thanking Mrs. Porterfield, who has served as our high school's Mathcounts coach for many years. She made it her mission to afford all of her students an opportunity to succeed in math. I am grateful for the two years I spent in her class, where she introduced me to many middle and high school math contests, which helped fuel my passion for mathematics. Thank you for the time and effort you spent coaching Mathcounts early in the mornings while the sky was still dark; I appreciate the countless hours you put into organizing contests.

I would also like to thank my creative, artistic, 7-year old sister Emily for giving me ideas of the front cover design.

I would like to thank the following contests for providing me with their problems:

The MATHCOUNTS Series is a premier middle school math competition that aims to help students develop a passion for mathematics, as well as build confidence and critical thinking skills.

The American Math Competitions (AMC 10, AMC 12) is a series of examinations for middle and high school students that build problem-solving skills and mathematics knowledge.

Contents

Chapter 1

An Introduction

Geometry is a branch of mathematics focused on the study of the sizes, positions, angles, and dimensions of an object. The word geometry originates from the Greek word geometria - to measure the Earth. In ancient Greece, the scholar Eratosthenes used geometry to be the first person to measure the diameter of the Earth. Since being founded by Euclid in Alexandria, Egypt in 300 BC, geometry has been applied throughout history to build grandiose structures such as the Colosseum in Rome, the Taj Mahal in India, and even modern skyscrapers such as the Burj Khalifa in Dubai.

If you look hard enough, you will see that geometry occurs all around you. From the beautiful hexagonal prisms found in snowflakes to the fractals found in Romanesco Broccoli, which form beautiful spirals that follow the Fibonacci sequence, nature is full of geometry. In this book, you will uncover the secrets and gain a deep appreciation of geometry.

Chapter 2

Fundamental Knowledge

2.1 Lines and Angles

2.1.1 Angles from intersecting lines

Angles are generated through the crossing of two lines.

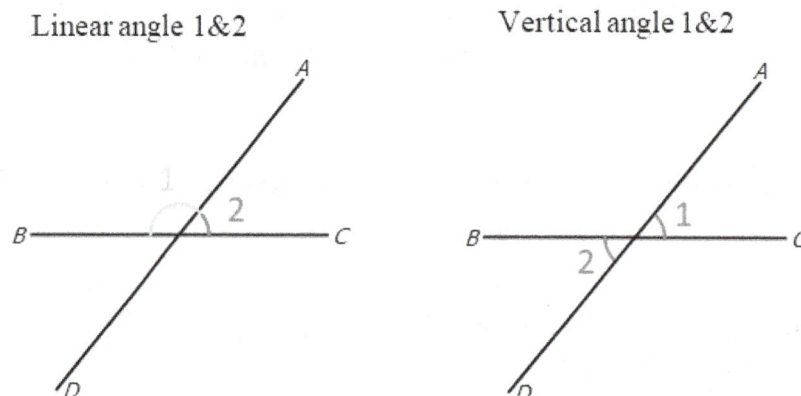

(1) Angles 1 and 2 (left diagram) are linear angles and they are supplementary, meaning that the sum of the angles is 180°.

(2) Angles 1 and 2 (right diagram) are called vertical angles. Vertical angles are congruent.

2.1.2 Angles from parallel lines

Three types of angle pairs are created when two parallel lines are crossed by a transversal line: corresponding angles, alternate interior angles, and alternate exterior angles. The angles in these angle pairs are congruent and its converse is also true.

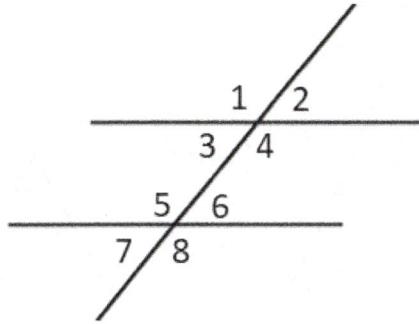

(1) Corresponding angles: 1&5, 2&6, 3&7, 4&8

(2) Alternate interior angles: 3&6, 4&5

(3) Alternate exterior angles: 1&8, 2&7

2.1.3 Angles from perpendicular lines

Perpendicular lines create 90° angles.

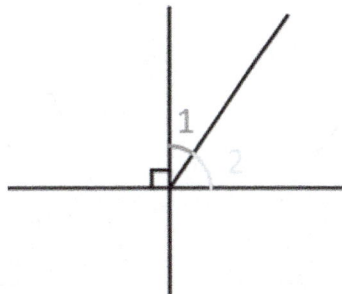

(1) $\angle 1 + \angle 2 = 90°$

2.2 Triangles

2.2.1 Types of triangles

(a) **Acute, right and obtuse triangles**

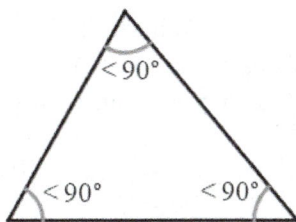

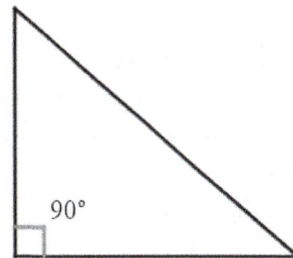

Acute triangle

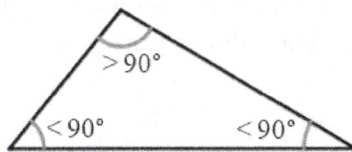

Obtuse triangle

Right triangle

7

(b) <u>triangle and equilateral triangle</u>

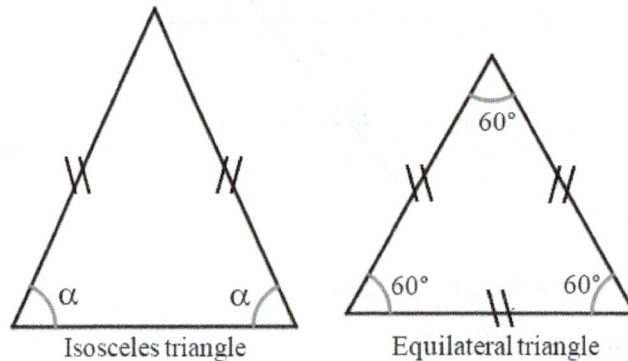

Isosceles triangle   Equilateral triangle

2.2.2   Altitudes, medians, midlines, angle and perpendicular bisectors

(a) **Altitudes of a triangle**

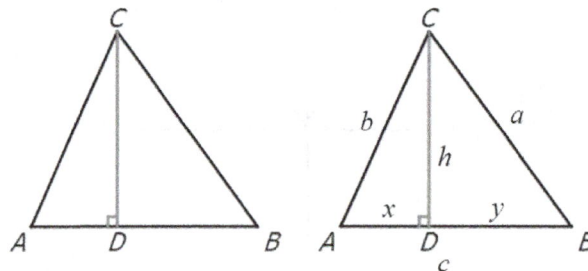

(1) In $\triangle ABC$, CD is the altitude to side AB.

(2) The area of $\triangle ABC$ is equal to $ch/2$.

(3) Given a, b and c as the side lengths of a triangle, **Heron's formula** can be used to calculate the area. $A = \sqrt{s(s-a)(s-b)(s-c)}$, where $s = (a+b+c)/2$. So the altitude $h = 2A/c$.

(4) Given the side lengths of a triangle as a, b and c, using the **Pythagorean Theorem** yields $a^2 - b^2 = y^2 - x^2 = c(y-x)$ and $x+y = c$. Thus $y = (a^2 + c^2 - b^2)/2c$ and $h = \sqrt{a^2 - y^2}$.

(5) Given the side lengths of a triangle as a, b and c, then we know $h = b\sin A$. If the measure of angle A is not given, however, we must use the **Law of cosines** to find cos A, which results in $\cos A = (b^2 + c^2 - a^2)/(2bc)$. Then the identity $(\sin A)^2 + (\cos A)^2 = 1$ can be used to solve for sin A.

(b) <u>Medians of a triangle</u>

8

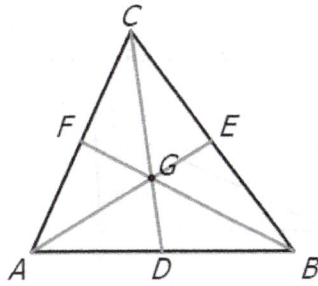

(1) In $\triangle ABC$, if D is the midpoint of segment AB, then CD would the median the side AB.

(2) The three medians of the triangle , CD, AE and BF, meet at a the centroid G, which is the center of gravity.

(3) The distance from the centroid to each vertex is $2/3$ the length of each median. $CG = 2GD$, $AG = 2GE$ and $BG = 2GF$.

(4) The area of each small triangle is $1/6$ of the area of $\triangle ABC$.

(5) The **Median Length Theorem**: $2BD^2 = AB^2 + BC^2 - AD^2 - BD^2$

(c) Midlines of a triangle

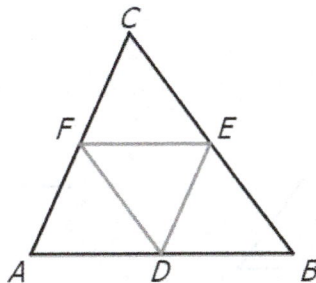

(1) Lines DE, EF and FD are midlines of $\triangle ABC$ when D, E and F are midpoints of side AB, BC and CA, respectively.

(2) $FE//AB$, $DE//AC$ and $DF//BC$.

(3) All of the triangles in the diagram are similar to each other.

(4) The area of each small triangle is $1/6$ of the area of $\triangle ABC$.

(d) Angle bisectors of a triangle

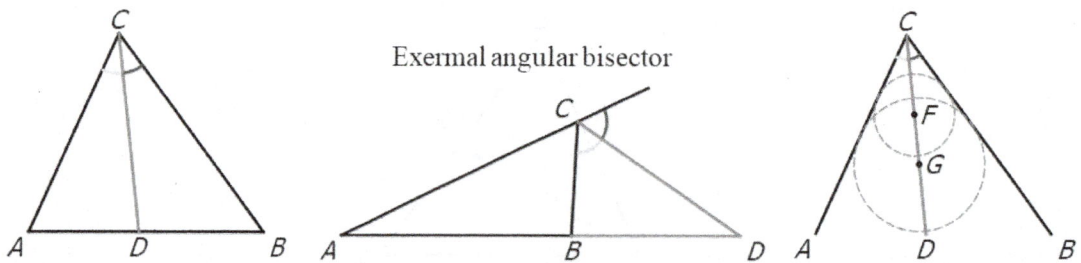

Exernal angular bisector

(1) CD is the angular bisector of $\angle BCA$ when $\angle BCD = \angle ACD$.

(2) The **Angular Bisector Theorem** says that if CD is the angle bisector of $\angle ACB$, then $AC/BC = AD/BD$ or $AC/AD = BC/BD$.

(3) The **Angular Bisector Length Theorem**: $CD^2 = AC * BC - AD * BD$.

(4) The **External Angular Bisector Theorem**: $AC/BC = AD/BD$ or $AC/AD = BC/BD$.

(5) The center of a circle tangent to both legs of an angle is on its angular bisector line.

(e) Perpendicular bisector of a triangle

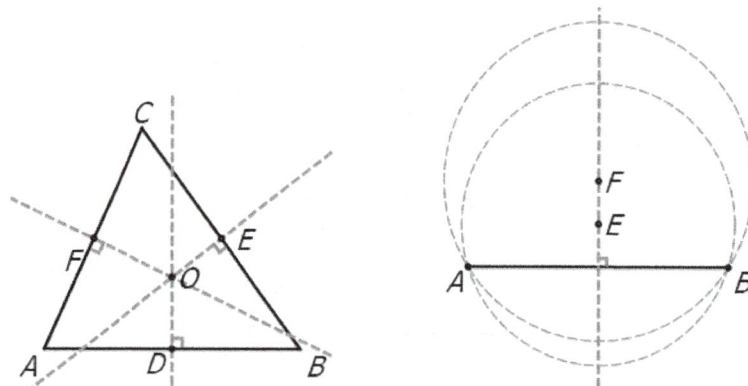

(1) The perpendicular bisector to AD is a line perpendicular to the segment (AB) that passes through its midpoint (D).

(2) Three perpendicular bisectors of a triangle meet at one point (O), which is the center of its circumcircle.

(3) The centers (e.g. E, F) of circles having segment AB as a cord is on the perpendicular bisector of AB.

10

2.2.3 Isosceles and equilateral triangles

(a) Isosceles triangles

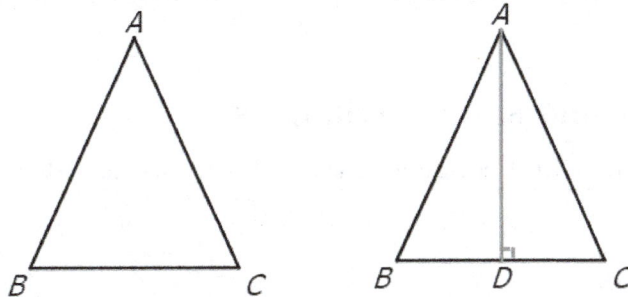

(1) $\triangle ABC$ is an isosceles triangle $\Longleftrightarrow AB = AC \Longleftrightarrow \angle B = \angle C$.

(2) In the isosceles $\triangle ABC$, the altitude AD of base side BC is also the median and angular bisector of $\angle A$.

(3) The length of the altitude can be found by the equation $AD^2 = AC^2 - CD^2$.

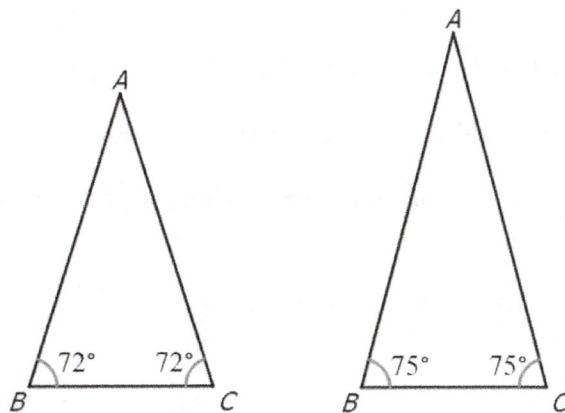

(4) An isosceles $\triangle ABC$ with $\angle B = \angle C = 72°$ has $AC/BC = (1 + \sqrt{5})/2$.

(5) An $\triangle ABC$ with $\angle B = \angle C = 75°$ has $AC/BC = (\sqrt{2} + \sqrt{6})/2$.

(b) Equilateral triangles

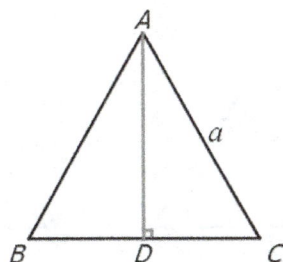

(1) $\triangle ABC$ is an equilateral triangle $\Rightarrow AB = AC = BC \Leftrightarrow \angle A = \angle B = \angle C = 60°$.

(2) An equilateral triangle ABC with side length of a has the area of $\sqrt{3}a^2/4$.

(3) The altitude of an equilateral triangle ABC with side length of a is $\sqrt{3}a/2$.

2.2.4 Congruent and similar triangles

(a) Determining similar and congruent triangles by using sides (S) and angles (A)

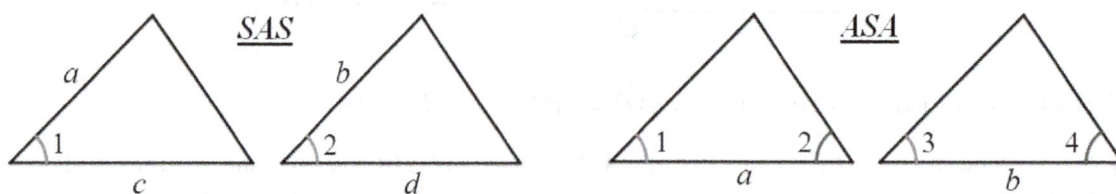

(1) The symbol for congruent triangle is $\cong$, e.g. $\triangle ABC \cong \triangle DEF$.

(2) SAS Congruence: if $a = b$, $c = d$ and $\angle 1 = \angle 2$.

(3) ASA Congruence: if $\angle 1 = \angle 3$, $a = b$ and $\angle 2 = \angle 4$.

(4) SSS Congruence: all three sides are equal for two triangles.

(5) SSA Congruence: $a = b$, $c = d$ and corresponding angle (corresponding to side c and d) $\angle 1 = \angle 2$.

(6) HL for right triangles: $a = b$ and $c = d$.

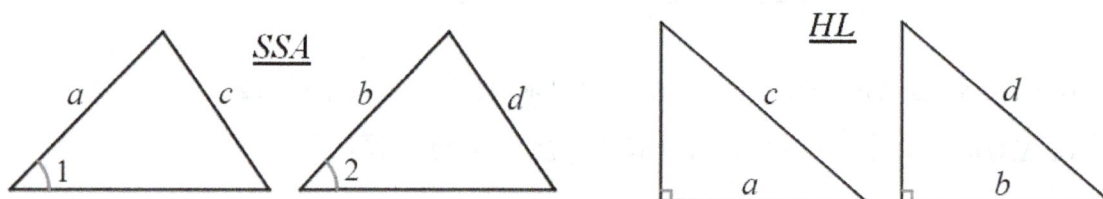

(b) Similar triangles

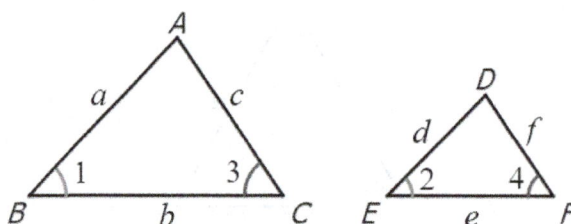

(1) The symbol for similar triangle is $\sim$, e.g. $\triangle ABC \sim \triangle DEF$.

(2) $\triangle ABC$ is an equilateral triangle $\Longleftrightarrow AB = AC = BC \Longleftrightarrow \angle A = \angle B = \angle C = 60°$.

(3) $\triangle ABC \sim \triangle DEF \Longleftrightarrow \angle 1 = \angle 2, \angle 3 = \angle 4$ and $a/d = b/e = c/f$.

(4) $\triangle ABC \sim \triangle DEF \Longleftrightarrow S_{\triangle ABC}/S_{\triangle DEF} = (a/d)^2 = (b/e)^2 = (c/f)^2$.

(5) Common similar triangles with parallel lines ($\triangle ABC \sim \triangle AEF$):

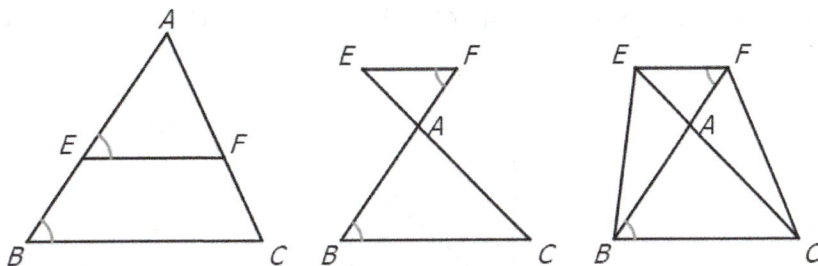

(6) Common similar triangles with shared angles ($\triangle ABC \sim \triangle AEF$):

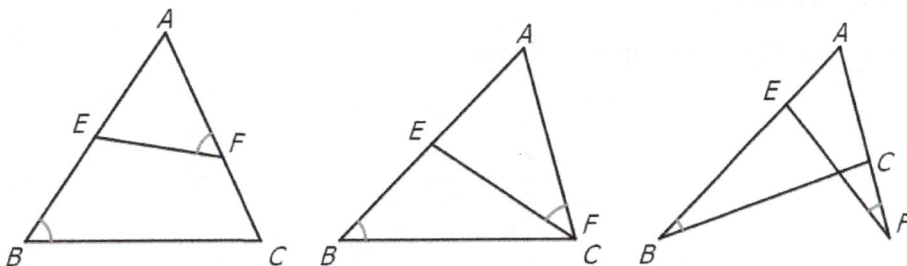

(7) Common similar triangles from right triangles ($\triangle ABC \sim \triangle AEF$):

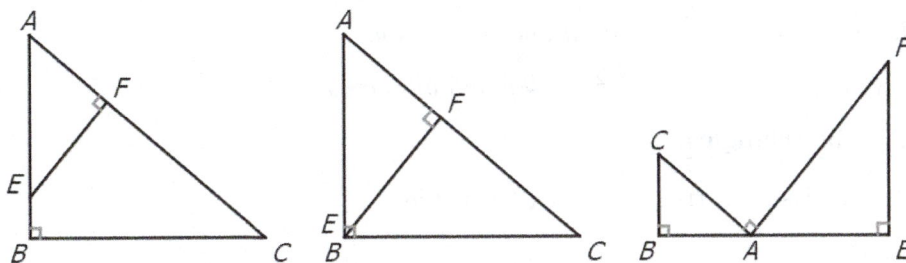

2.2.5 The Pythagorean theorem

(a) **The Pythagorean Theorem:** In a right triangle $\triangle ABC$, $\angle C = 90° \rightarrow a^2 + b^2 = c^2$, where a and b are the length of two legs and c is the length of the hypotenuse.

13

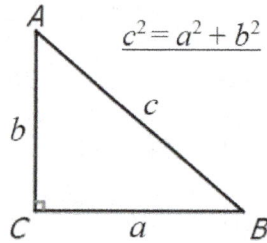

$$c^2 = a^2 + b^2$$

(b) The converse of the Pythagorean Theorem: $a^2 + b^2 = c^2 \to \angle C = 90°$ and $\triangle ABC$ is a right triangle.

(c) Median of a right triangle: In $\triangle ABC$, $CM = AM = BM \to \angle C = 90°$ (right triangle).

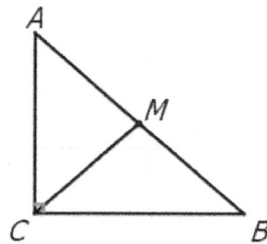

2.2.6 Right triangles

(a) Extensive relations among side-lengths

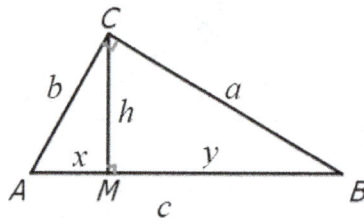

(1) The Pythagorean Theorem: $a^2 + b^2 = c^2$.

(2) Altitude relationships: $a * b = c * h$ and $h^2 = x * y$.

(3) Similar triangle relationships: $a^2 = c * y$ and $b^2 = c * x$.

(b) Special right triangles

(1) 45-45-90 isosceles right triangle: $c = \sqrt{2}b = \sqrt{2}a$.

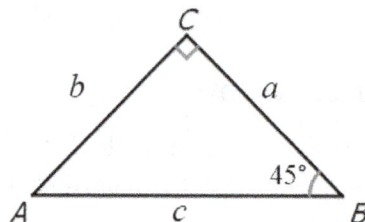

(2) 30-60-90 right triangle with one angle of 30°(or 60°): $c = 2b$ and $a = \sqrt{3}b$.

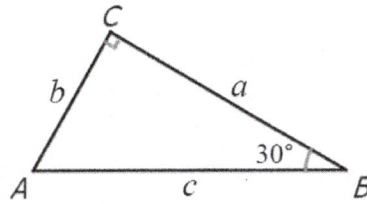

(c) Inscribed circle of a right triangle

The radius of an inscribed circle of a right triangle can be found by $r = (a + b - c)/2$.

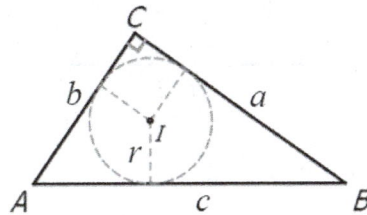

(d) Circumcircle of a right triangle

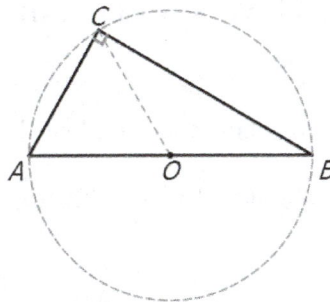

(1) The center of the circumcircle of a right triangle is the midpoint of its hypotenuse.

(2) The hypotenuse is the diameter of the circumcircle.

2.2.7 Inequalities in a triangle

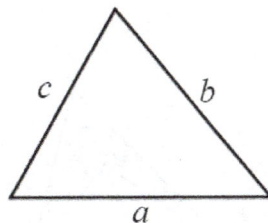

(a) The triangle inequality theorem

15

In any triangle with sides length a, b and c, the following inequalities are true:

(1) $a + b > c$, $a + c > b$ and $b + c > a$.

(2) $|a - b| < c$, $|b - c| < a$ and $|a - c| < b$.

2.2.8 Area of a triangle

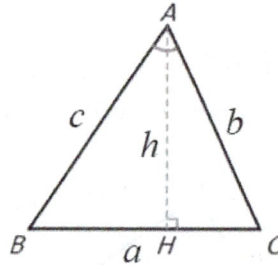

(1) If the altitude is known, $S = a * h / 2$.

(2) $S = (b * c * \sin A)/2$.

(3) **Heron's formula**: $S = \sqrt{s(s-a)(s-b)(s-c)}$, where $s = (a + b + c)/2$.

2.2.9 The Area method, Ceva's Theorem, and Menelaus' Theorem

(a) The Area method

The following are useful patterns to relate area and length of segments:

(1) Scenarios involving parallel lines: $AD // BC \longrightarrow S_{\triangle PBD} = S_{\triangle PAC}$.

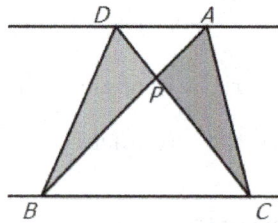

(2) Scenarios involving shared sides: $S_{\triangle ADB}/S_{\triangle ADC} = BP/PC$.

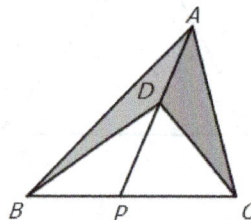

(3) Quadrilaterals with shared sides (e.g. AC): $S_{\triangle ACD}/S_{\triangle ABC} = DE/EB$.

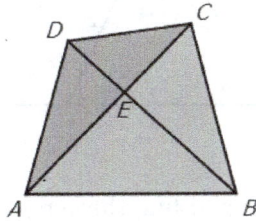

(4) Triangles with shared bases (e.g. BP): $S_{\triangle ABP}/S_{\triangle DBP} = AC/DC$.

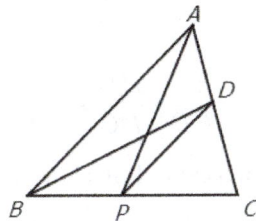

(b) **Ceva's Theorem**

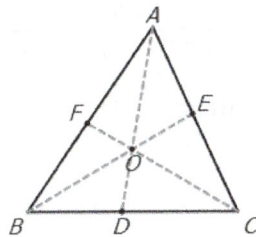

(1) Given a triangle ABC with the lines AO, BO, and CO drawn from the vertices to a common point O to meet the opposite sides at D, E and F, the following is true:

$$\frac{AF}{FB} \cdot \frac{BD}{DC} \cdot \frac{CE}{EA} = 1$$

(2) The converse of Ceva's theorem: If points D, E and F are on located sides BC, CA and AB, respectively, such that

$$\frac{AF}{FB} \cdot \frac{BD}{DC} \cdot \frac{CE}{EA} = 1$$

Then AD, BE and CF must be concurrent, meaning that they intersect at a common point.

(c) **Menelaus' Theorem**

17

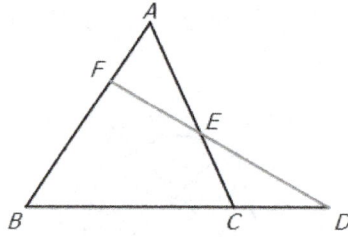

(1) Given a triangle ABC and transversal line that crosses BC, AC and AB at points D, E and F, respectively, then

$$\frac{AF}{FB} \cdot \frac{BD}{DC} \cdot \frac{CE}{EA} = 1$$

(2) The converse of Menelaus' Theorem: If points D, E and F are on side BC, CA and AB, respectively, such that

$$\frac{AF}{FB} \cdot \frac{BD}{DC} \cdot \frac{CE}{EA} = 1$$

Then D, E and F are collinear, meaning that they lie on the same line.

2.2.10 The Law of Sines and Cosines

(a) The Law of Sines

(1) In $\triangle ABC$,

$$\frac{a}{sin(A)} = \frac{b}{sin(B)} = \frac{c}{sin(C)} = 2r$$

where r denotes the radius of its circumcircle.

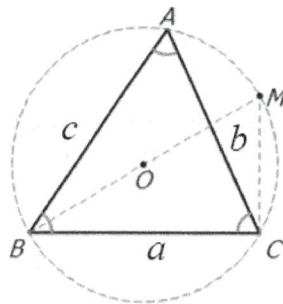

(b) The Law of Cosines

(1) The cosine of an angle: $\cos C = \frac{a^2+b^2-c^2}{2ab}$

(2) The length of one side: $c^2 = a^2 + b^2 - 2ab\cos C$

18

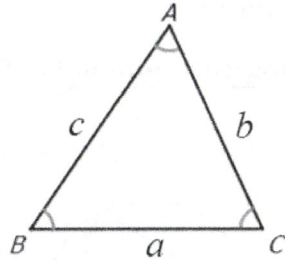

2.3 Circles

2.3.1 Circumference and area

(a) **Circumference and arc length**

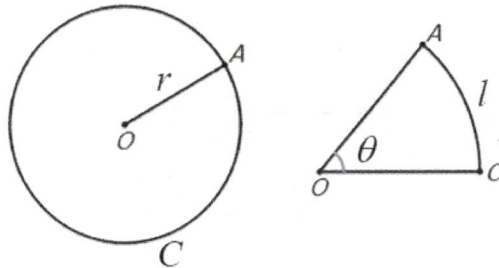

(1) The circumference of a circle with radius of r: $C = 2\pi r$

(2) The length of arc with radius of r: $l = \frac{\theta r}{2\pi}$

(b) **The area of a circle and a sector**

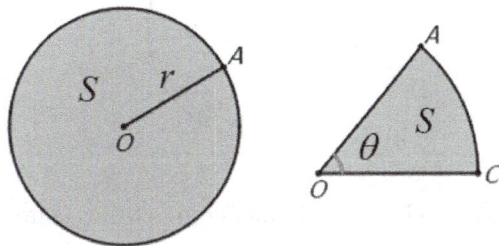

(1) The area of a circle with radius of r: $S = \pi r^2$

(2) The area of area with radius of r: $S = \theta r^2 / 2$

2.3.2 The diameter of a circle

(a) The inscribed angle containing the endpoints of the diameter is 90°:

BC is a diameter $\Rightarrow \angle BAC = 90°$

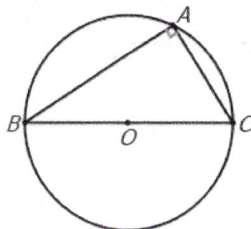

(b) A chord perpendicular to the diameter is bisected:

Diameter $BC \perp AD \Leftrightarrow AE = DE$

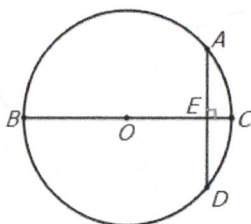

2.3.3 Tangent lines to circles

(a) A line tangent to a circle touches the circle at a single point

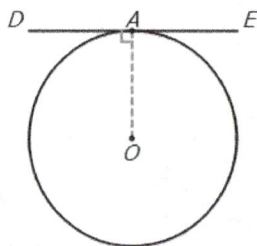

(1) Given a line tangent to a circle, you should connect the center of the circle to the tangent point.

(2) $OA = r$ and $OA \perp DE$

(b) Two tangent lines to a circle from one external point

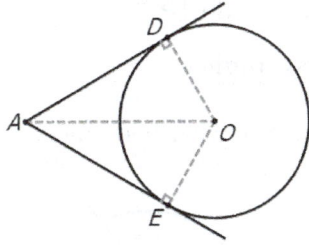

(1) $\triangle ADO \cong \triangle AEO$

(2) $OD \perp AD$ and $OE \perp AE$

(3) $AD = AE$ and $OA^2 = AD^2 + OD^2$

(c) **External tangent lines of two circles**

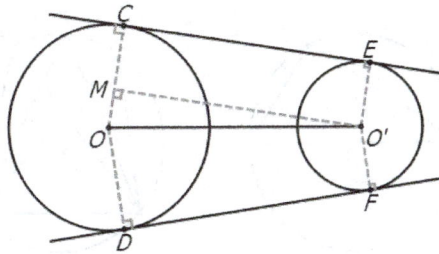

(1) $OC \perp CE$ and $O'E \perp CE$

(2) $CE = DF$ and $OO'^2 = CE^2 + (OC - O'E)^2$

(d) **Internal tangent lines of two circles**

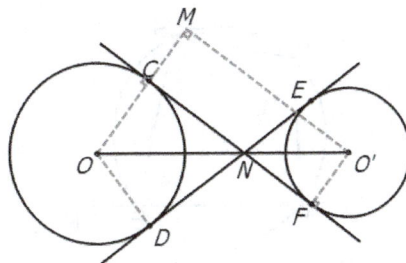

(1) $OC \perp CF$ and $O'F \perp CF$

(2) $ON/NO' = OC/O'F = R/r$

(3) $CE = DE$ and $OO'^2 = CF^2 + (OC + O'F)^2$

2.3.4 Inscribed angles in a circle

(a) Central angles and inscribed angles

The central angle of a cord is twice the inscribed angle: $\angle DOE = 2\angle A$

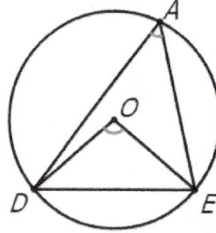

(b) Inscribed angle and cords

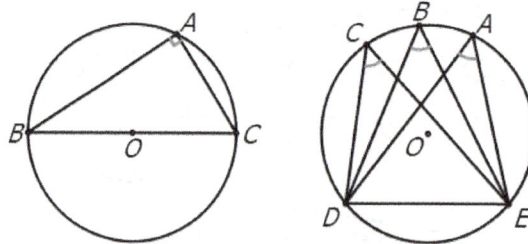

(1) The corresponding inscribed angle of a diameter is $90°(\angle A = 90°)$.

(2) All inscribed angles corresponding to the same cord are equal: $\angle A = \angle B = \angle C$

(c) Four concyclic points and external angle

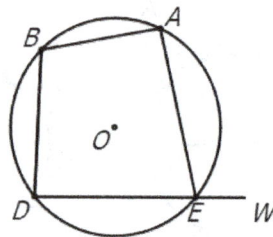

(1) Supplementary angle pairs: $\angle A + \angle D = 180°$ and $\angle B + \angle E = 180°$

(2) Equal external angle: $\angle AEW = \angle B$

(3) Center of the circumcircle is the crossing point of two perpendicular bisectors of two chords (non-parallel).

2.3.5   Inscribed and circumscribed circles

(a) Let the radius of the circumcircle of $\triangle ABC$ be R

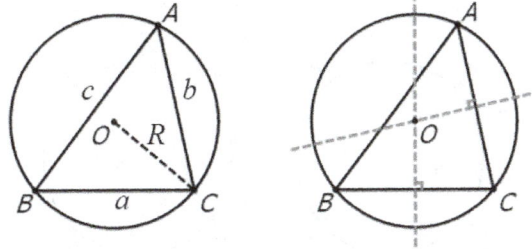

(1) The area of the triangle ABC is $S_{\triangle ABC} = (abc)/(4R)$

(2) The Law of Sines yields
$$\frac{a}{sin(A)} = \frac{b}{sin(B)} = \frac{c}{sin(C)} = 2R$$

(3) The center of the circumcircle is the intersection point of the perpendicular bisectors of $\triangle ABC$.

(b) The inscribed circle

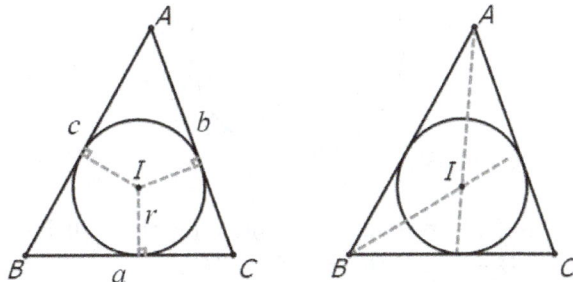

(1) The area of the triangle ABC is $S_{\triangle ABC} = rs$, where $s = (a + b + c)/2$.

(2) The center of the inscribed circle is the intersection point of the angular bisectors of $\triangle ABC$.

2.3.6   The Power of a Point Theorem

(a) Two intersecting chords

If the chords AC and BD cross at a point P, then $PA * PC = PB * PD$.

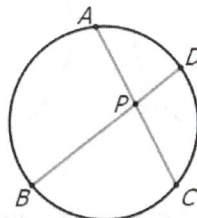

(b) Two rays from one external point

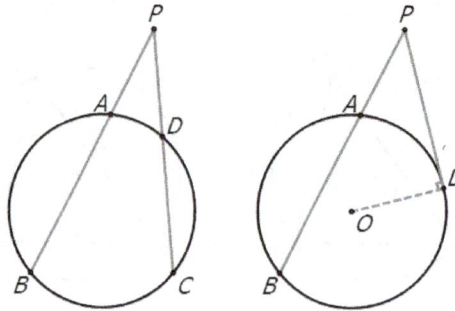

(1) $PA * PB = PD * PC$

(2) $PA * PB = PD^2$, where PD is tangent to the circle.

2.3.7 Ptolemy's Theorem

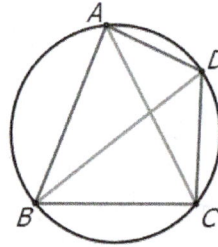

(1) If the points A, B, C and D are concyclic, meaning they all lie on a circle, then Ptolemy's Theorems yields $AB * CD + AD * BC = AC * BD$.

(2) The converse of Ptolemy's theorem is also true: If $ABCD$ is a quadrilateral such that $AB * CD + AD * BC = AC * BD$, then the points A, B, C and D are concyclic.

2.3.8 Concyclic (or cocyclic) points

To show that four points are concyclic, analyze the angles or the side-lengths:

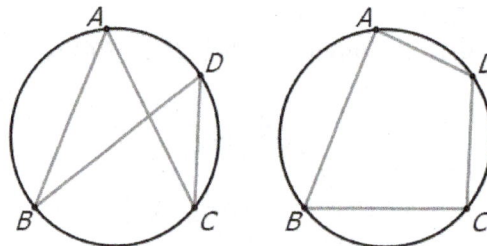

(1) Angles with endpoints on the same arc segment are congruent: $\angle A = \angle D$ or $\angle B = \angle C$

24

(2) Opposite angles are supplementary: $\angle A + \angle C = 180°$ or $\angle B + \angle D = 180°$

(3) The converse of Ptolemy's Theorem

(4) The converse of the Power of a Point Theorem

2.4 Polygons

2.4.1 Parallelograms

A parallelogram is a special type of quadrilateral with two pairs of parallel sides (e.g. $AB//CD$ and $AD//BC$). Rectangles and rhombus' are special parallelograms with congruent angles and sides, respectively.

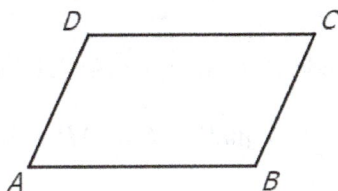

(a) Edge-length and angles

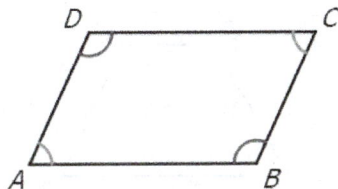

(1) $AB = CD$ and $AD = BC$

(2) $\angle A = \angle C$ and $\angle B = \angle D$

(3) $\angle A + \angle B = 180°$ and $\angle B + \angle C = 180°$

(b) Diagonals

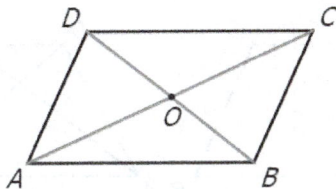

(1) Diagonals AC and BD bisect each other: $OA = OC$ and $OB = OD$.

(2) The sum of the squares of the diagonals is equal to the sum of the squares of all four sides: $AC^2 + BD^2 = AB^2 + BC^2 + CD^2 + DA^2$.

(3) Point O is the center of symmetry of the parallelogram, meaning that any line that passes though point O divides the parallelogram into two congruent shapes.

(c) **Area of parallelogram**

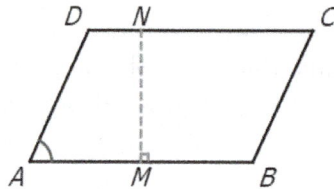

(1) Given MN is the altitude to side AB, the area $S = AB * MN$

(2) The area can also be found by the equation $S = AB * AD * \sin A$

2.4.2 Trapezoids

(a) A **Trapezoid** is a quadrilateral with one pair of parallel sides.

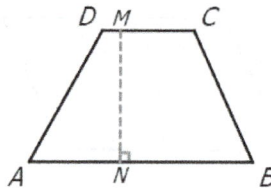

(1) $AB // CD$

(2) $\angle A + \angle D = 180°$ and $\angle B + \angle C = 180°$

(3) The area of the trapezoid is $S = (AB + CD) * MN/2$, where MN is an altitude.

(b) **Segment lengths in trapezoid**

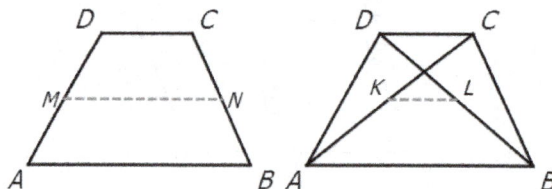

(1) The length of the median of a trapezoid parallel to the bases: $MN = (AB + CD)/2$

26

(2) The length of the segment connecting midpoints of the diagonals: $KL = (AB - CD)/2$

(3) Given that segment EF passes through the intersecting point of diagonals and is parallel to the base: $EF = AB * CD/(AB + CD)$

(4) Given that segment GH connects the two midpoints of AB and CD: $4GH^2 = 2(AD^2 + BC^2) - (AB - CD)^2$

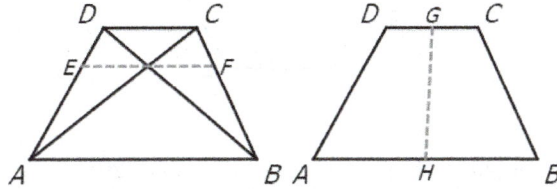

(c) Areas in parallelogram

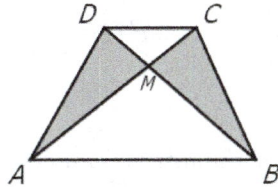

(1) $S_{\triangle AMD} = S_{\triangle BMC}$

(2) $S_{\triangle AMD} * S_{\triangle BMC} = S_{\triangle AMB} * S_{\triangle CMD}$

2.4.3 Pentagons

Pentagon is a five-sided polygon (e.g. concave pentagon, normal pentagon).

(a) Normal pentagons

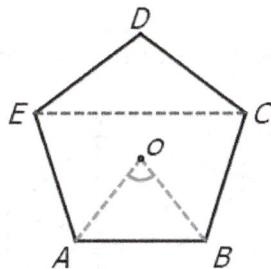

(1) The sum of the internal angles is 540°.

(2) Each internal angle has a measure of 108°, and each exterior angle has a measure of 72°.

(3) The ratio of the length of a diagonal (EC) to the side length (AB) is $(1 + \sqrt{5})/2$.

(4) The area of a normal pentagon can be found by the equation $S = 5OA^2 * \sin 72°/2$.

27

2.4.4   Hexagons and other polygons

(a) Normal Hexagons

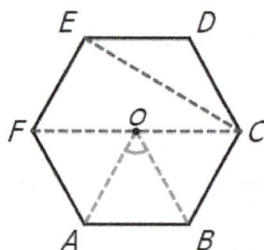

(1) In a normal hexagon, the sum of the internal angles is 720°.

(2) Each internal angle has a measure of 120° and each exterior angle has a measure of 60°.

(3) Given a side-length of a, the length of diagonals CE and CF are $\sqrt{3}a$ and $2a$, respectively.

(4) The area of a normal hexagon is $S = 3\sqrt{3}a^2/2$

(b) Other Polygons

(1) In an n-sided polygon, the sum of the internal angles $180\times(n\text{-}2)$ degrees.

(2) The sum of the exterior angles of any convex polygon is 360°.

(3) In a normal polygon with n sides, each external angle has a measure of $360/n$ degrees and each internal angle has a measure of $180\text{-}(360/n)$ degrees.

2.5   Volume

2.5.1   Cubes and rectangular prisms

(a) Cubes

A cube has 6 square faces, 12 edges and 8 vertices. Given the edge-length of a,

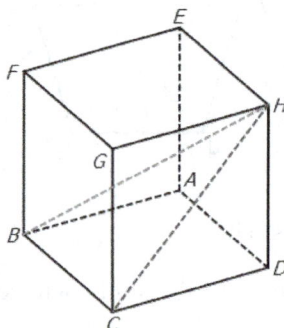

(1) The volume of a cube is $V = a^3$

(2) The surface area of a cube is $S = 6a^2$

(3) The length of the diagonal of a cube connecting opposite corners is $BH = \sqrt{3}a$ and the diagonal of a face is $CH = \sqrt{2}a$.

(b) Rectangular prism

A rectangular prism has 6 rectangular faces, 12 edges and 8 vertices.

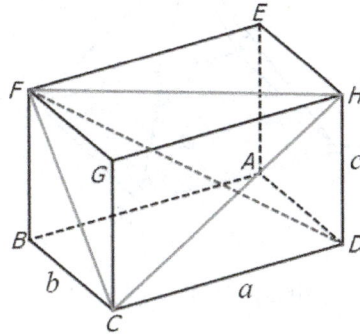

(1) Given edge-lengths of a, b and c, the volume is $V = abc$.

(2) The surface area is $S = 2(ab + bc + ca)$.

(3) The length of diagonals: $FD = \sqrt{a^2 + b^2 + c^2}$, $FC = \sqrt{b^2 + c^2}$, $CH = \sqrt{a^2 + c^2}$ and $FH = \sqrt{a^2 + b^2}$.

2.5.2 Triangular pyramids and square pyramids

(a) Regular triangular pyramids

A regular triangular pyramid is a tetrahedral that consists of four equilateral triangle faces.

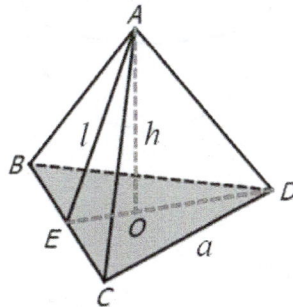

(1) The volume of a generic pyramid is $V = hB/3$, where B is the area of base triangle and h is the altitude.

(2) The altitude of a regular triangular pyramid with an edge-length of a is $h = \sqrt{6}a/3$.

(3) The volume of the tetrahedral is $V = \sqrt{2}a^3/12$.

(4) The slant height is $l = \sqrt{3}a/2$.

(b) Regular square pyramids

A regular square pyramid has equal side lengths and consists of four equilateral triangle sides and a square base.

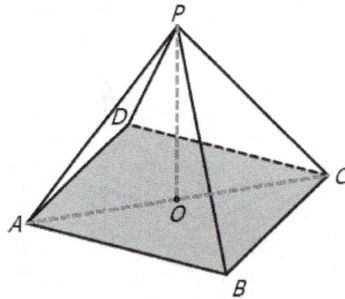

(1) Given an edge-length of a, the altitude is $h = \sqrt{2}a/2$.

(2) The volume of the regular square pyramid is $V = \sqrt{2}a^3/6$.

(3) The slant height is $l = \sqrt{3}a/2$.

(4) $\triangle PAC$ and $\triangle PBD$ are isosceles right triangles.

2.5.3 Spheres

A sphere with the radius of r:

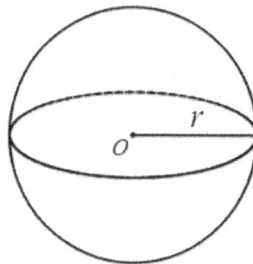

(1) The surface area of a sphere: $S = 4\pi r^2$

(2) The volume of a sphere: $V = 4\pi r^3/3$

(3) The intersection of a plane with a sphere results in a circle, which has the radius of r if the plane passes through the center of the circle.

30

2.5.4 Cones

The radius of the base is r, the height is h and the side length is l:

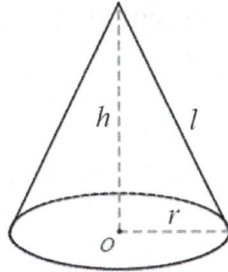

(1) The total surface area of the cone is $S = \pi r^2 + \pi r l$.

(2) The volume of the cone is $V = \pi r^2 h / 3$.

(3) The slant height can be found by the equation $l^2 = r^2 + h^2$.

2.5.5 Frustums

(a) Frustums of a right circular cone:

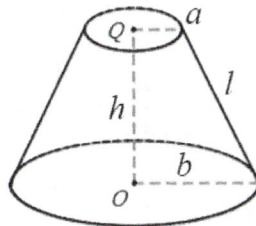

(1) The total surface area of the cone: $S = \pi a^2 + \pi b^2 + \pi(a + b)l$

(2) The volume of the cone is $V = \pi h(a^2 + ab + b^2)/3$.

(3) The slant height: $l^2 = (b - a)^2 + h^2$

(b) Frustum from a pyramid

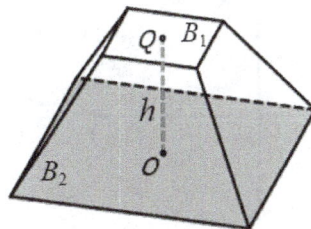

(1) The volume of a generic frustum from a pyramid: $V = h(B_1 + \sqrt{B_1 B_2} + B_2)/3$

(2) The volume of a frustum with squares as both bases: $V = \pi h(a^2 + ab + b^2)/3$

(3) The slant height can be calculated using the equation: $l^2 = (a - b)^2/4 + h^2$

31

2.6 Analytical Geometry

2.6.1 Coordinate system basics

(a) **Each point in a 2D space is represented by a pair of coordinates (x, y).**

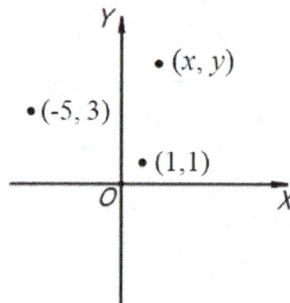

(b) **A collection of points are used to represent more complex shapes**

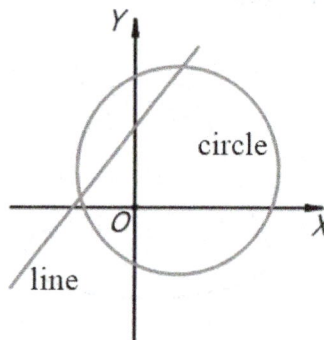

(c) **Explicit/implicit functions to represent geometric shapes**

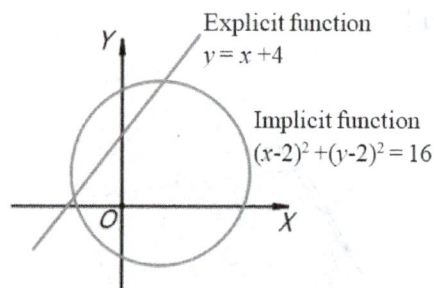

(1) An explicit function has the form: $y = f(x)$, e.g., $y = 3x + 2$, $y = 2x^2 - 1$

(2) An implicit function has the form: $f(x, y) = c$, e.g., $x^2 + y^2 = 1$, $3x^2 + 2y^2 = 6$

(d) **The solution of the equation set represents the intersection points.**

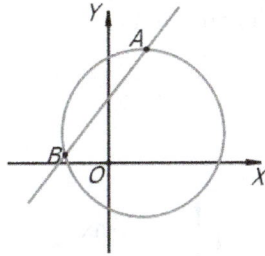

(1) The coordinates of intersecting points are equivalent to the solutions of the equation set.

(2) The number of intersecting points is equal to the number of roots of the equation set.

2.6.2 Lines in coordinate system

(a) Types of equations for a line

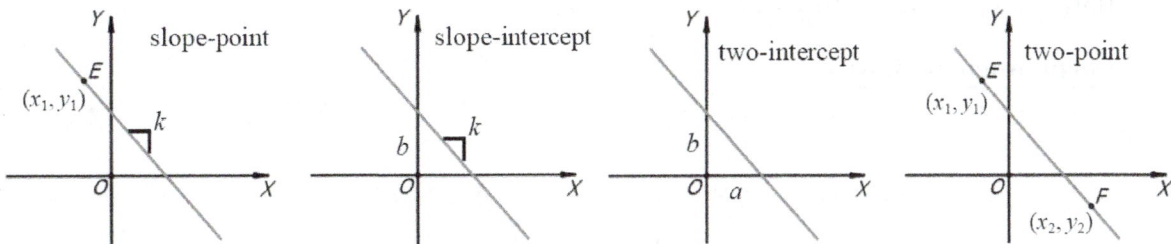

(1) Standard form: $ax + by = c$, where $a^2 + b^2 \neq 0$

(2) Slope-point form: $y - y_1 = k(x - x_1)$

(3) Slope-intercept form: $y = kx + b$

(4) Two-intercept form: $x/a + y/b = 1$

(5) Two-point form: $(y - y_1)/(x - x_1) = (y_2 - y_1)/(x_2 - x_1)$

(b) Calculating slopes

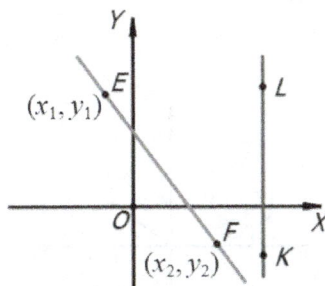

(1) $k = (y_2 - y_1)/(x_2 - x_1)$, if $x_2 \neq x_1$.

33

(2) If $x_1 = x_2$, then the line is parallel to y-axis.

(c) Parallel lines

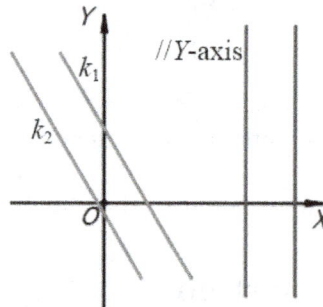

(1) Parallel lines have the same slope ($k_1 = k_2$) or

(2) Both lines are parallel to y-axis.

(d) Perpendicular lines

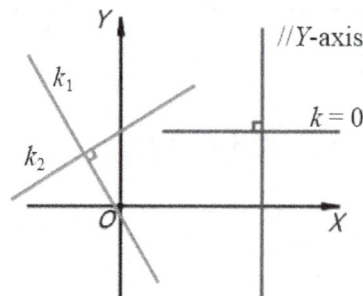

(1) The product of the slopes of perpendicular lines is -1 ($k1 * k2 = -1$) or

(2) If one line is parallel to y-axis, the other line is parallel to x-axis ($k = 0$).

2.6.3 Circles in coordinate system

(a) Circles centered at the origin

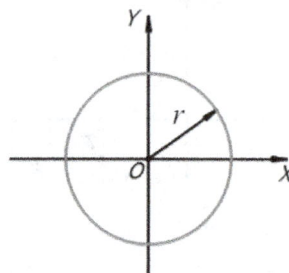

(1) The equation of a circle centered at the origin is $x^2 + y^2 = r^2$.

(2) The intercepts are $\pm r$ for both x-axis and y-axis.

(b) Circles with arbitrary centers

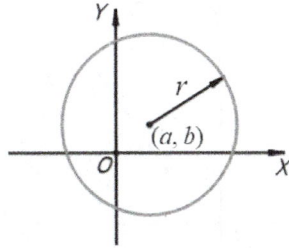

(1) The equation of a circle centered at (a, b) with a radius of r is $(x - a)^2 + (y - b)^2 = r^2$.

(2) The origin O is inside the circle if $a^2 + b^2 < r^2$.

(3) The origin O is outside the circle if $a^2 + b^2 > r^2$.

(c) Equation of a circle in general form

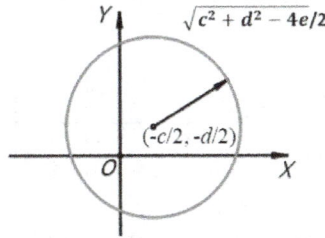

(1) General form of a circle: $x^2 + y^2 + cx + dy + e = 0$

(2) Convenient form: $(x + c/2)^2 + (y + d/2)^2 = (c^2 + d^2 - 4e)/4$

(3) The center of the circle is $(-c/2, -d/2)$ and radius is $\sqrt{c^2 + d^2 - 4e}/2$, where $c^2 + d^2 - 4e > 0$.

2.6.4 The distance formula

(a) Midpoint formula

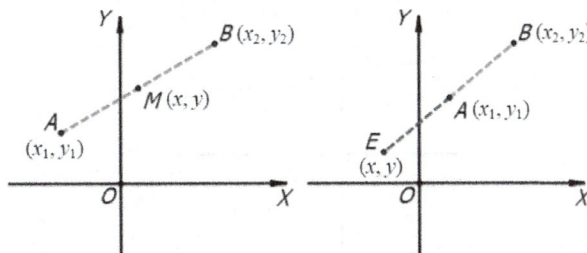

35

(1) Given two points (x_1, y_1) and (x_2, y_2), $x = (x_1 + x_2)/2$, $y = (y_1 + y_2)/2$

(2) If point (x_1, y_1) is reflected across point (x_2, y_2) to point (x, y) then $x = 2x_1 - x_2$ and $y = 2y_1 - y_2$.

(b) Distance between two points

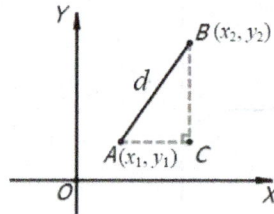

(1) From the **Pythagorean Theorem**, the distance between two points is
$d = \sqrt{(x_1 - x_2)^2 + (y_1 - y_2)^2}$.

(c) Distance between a point and a line

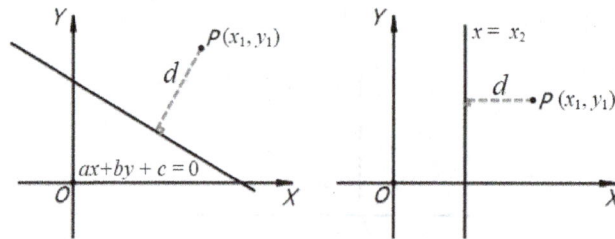

(1) The distance between the line $ax + by + c = 0$ and the point (x_1, y_1):

$$d = \frac{|ax_1 + by_1 + c|}{\sqrt{a^2 + b^2}}$$

(2) If the line is parallel to the y-axis ($x = x_2$), the distance to the point (x_1, y_1) is $d = |x_1 - x_2|$.

2.6.5 Geometry of Intersections

(a) Methods to calculate the length of segments from intersections:

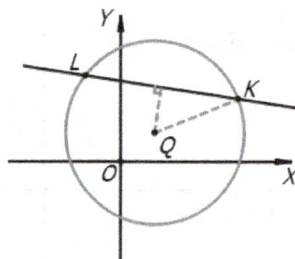

36

(1) Geometric method: calculate distance d, then use **Pythagorean Theorem**: $KL = 2\sqrt{r^2 - d^2}$

(2) Algebraic method: establish the equation set including the circle $x^2 + y^2 + cx + dy + e = 0$ and the line $y = kx + b$, use **Vieta Theorem** to calculate $|x_1 - x_2|$. The intersecting segment length $KL = \sqrt{k^2 + 1}|x_1 - x_2|$.

(b) Tangent lines to a circle:

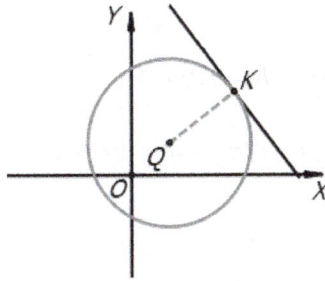

(1) Geometric method: the circle center to line distance $d = r$

(2) Algebraic method: if an equation set has only one root $\Longrightarrow$ discriminant of the quadratic equation Δ equals to zero.

2.6.6   Symmetry in coordinate system

(a) Symmetry about the origin

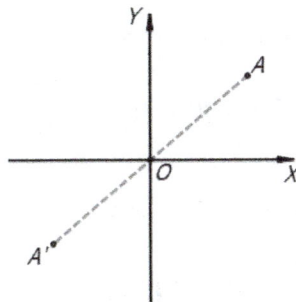

(1) Point $(x, y) \Rightarrow (-x, -y)$

(2) Explicit function $y = f(x) \Rightarrow y = -f(-x)$

(3) Implicit function $f(x, y) = 0 \Rightarrow f(-x, -y) = 0$

(b) Symmetry about point (a, b)

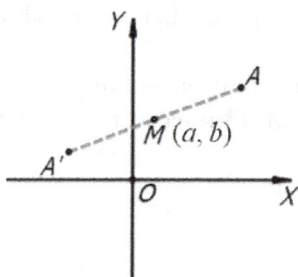

(1) Point $(x, y) \Rightarrow (2a - x, 2b - y)$

(2) Explicit function $y = f(x) \Rightarrow y = 2b - f(2a - x)$

(3) Implicit function $f(x, y) = 0 \Rightarrow f(2a - x, 2b - y) = 0$

(c) Symmetry over the x-axis

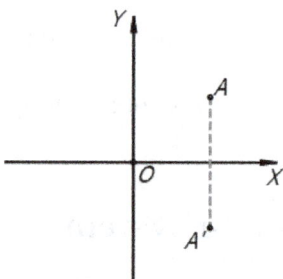

(1) Point $(x, y) \Rightarrow (x, -y))$

(2) Explicit function $y = f(x) \Rightarrow y = -f(x)$

(3) Implicit function $f(x, y) = 0 \Rightarrow f(x, -y) = 0$

(d) Symmetry over the y-axis

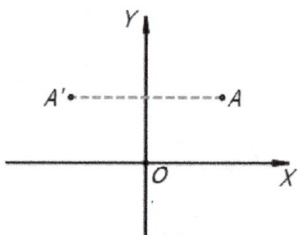

(1) Point $(x, y) \Rightarrow (-x, y)$

(2) Explicit function $y = f(x) \Rightarrow y = f(-x)$

(3) Implicit function $f(x, y) = 0 \Rightarrow f(-x, y) = 0$

(e) Reflection about the line y = x

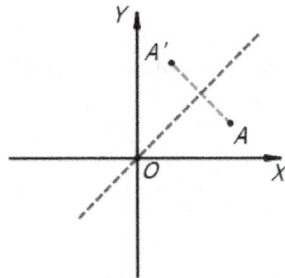

(1) Point $(x, y) \Rightarrow (y, x)$

(2) Explicit function $y = f(x) \Rightarrow y = (f(x))^{-1}$ (inverse function)

(3) Implicit function $f(x, y) = 0 \Rightarrow f(y, x) = 0$

(f) Reflection about the line y = -x

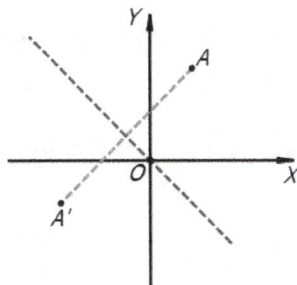

(1) Point $(x, y) \Rightarrow (-y, -x)$

(2) Explicit function $y = f(x) \Rightarrow y = -(f(-x))^{-1}$ (inverse function)

(3) Implicit function $f(x, y) = 0 \Rightarrow f(-y, -x) = 0$

2.6.7 Graphing absolute values

(a) Absolute value of x coordinate: $|x|$

$y = 2|x| + 1$

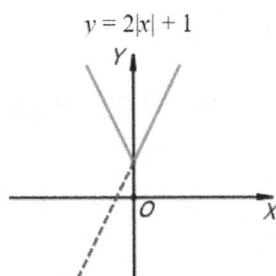

$(|x| - 2)^2 + (y - 2)^2 = 9$

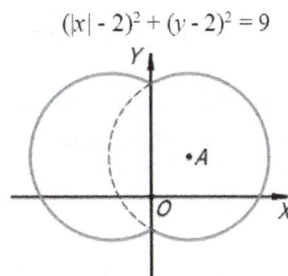

39

(1) The graph is symmetric about the y-axis (in red).

(2) For an explicit function $y = f(|x|)$, graph the $x > 0$ portion and reflect the graph across y-axis. Example: $y = 2|x| + 1$

(3) For an implicit function $f(|x|, y) = 0$, graph $x > 0$ portion and reflect the graph across y-axis. Example: $(|x| - 2)^2 + (y - 2)^2 = 9$

(b) **Absolute value of y coordinate: $|y|$**

$$y = |2x + 1| \qquad\qquad (x - 2)^2 + (|y| - 2)^2 = 9$$

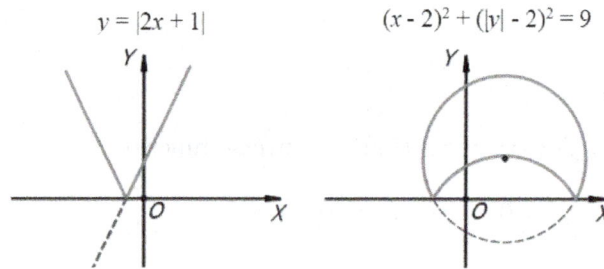

(1) The graph is symmetric about the x-axis.

(2) For an explicit function $y = |f(x)|$, graph $y = f(x)$ first and then reflect the portion below the x-axis above it.

(3) For an implicit function $f(x, |y|)$, graph $f(x, y) = 0$ first and then the reflect the portion below the x-axis above it.

(c) **Absolution value of both $|x|$ and $|y|$**

$$(|x| - 2)^2 + (|y| - 2)^2 = 9$$

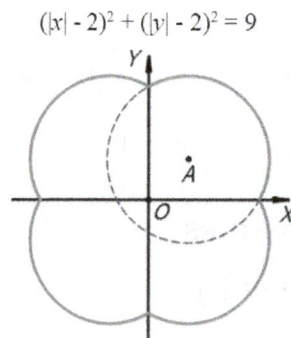

(1) The graph is symmetric about both x-axis and y-axis.

(2) For an implicit function $f(|x|, |y|)$, graph $f(x, y) = 0$ for $x > 0$ and $y > 0$ first and then reflect the shape across both x-axis and y-axis.

40

Chapter 3

Two-hundred-twenty-five Problems

It is time to apply the knowledge you've gained to solve some real geometry problems. We have assembled a comprehensive collection of 225 geometry problems taken from previous Mathcounts, AMC 8, and AMC 10/12 competitions. These problems will have detailed solutions and tips in order to guide you towards the most efficient solution we recommend using during a competition. Gear up and have a great ride!

Problem 1

Square $ABCD$ has sides of length 1 cm. Triangle CFE is an isosceles right triangle tangent to arc BD at G. Arc BD is a quarter circle with its center at A. What is the total area of the two shaded regions? (2005 Mathcounts National Team)

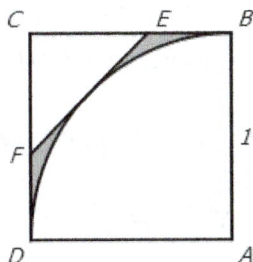

> **Tips:** *1. Focus on the diagonal of the square. 2. Use the properties of an isosceles right triangle.*

Solution 1:

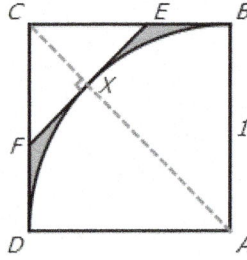

The shaded area is equal to the area of the square minus the area of the quarter circle and the area of the triangle.

Connect AC and let the intersection to the arc be point X. Clearly, we know that $AC = \sqrt{2}$ and $XC = \sqrt{2} - 1$. The isosceles right triangle $\triangle CEF$ has an area of $CX^2 = \left(\sqrt{2} - 1\right)^2 = 3 - 2\sqrt{2}$.

The area of the quarter circle is $\pi/4$, hence the shaded area is $1 - \pi/4 - (3 - 2\sqrt{2}) = 2\sqrt{2} - 2 - \frac{\pi}{4}$.

Solution 2:

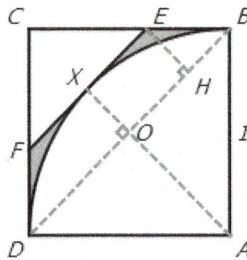

First calculate the area of isosceles trapezoid $EFDB$. From the diagram above, we can get

$$DB = \sqrt{2}OA = OB = OD = \frac{\sqrt{2}}{2}$$

$$OX = EH = BH = 1 - \sqrt{2}$$

So, $EF = DB - 2BH = 2\sqrt{2} - 2$. Thus the area of isosceles trapezoid $EFDB$ is

$$\frac{1}{2}(EF + DB)EH = 2\sqrt{2} - \frac{5}{2}$$

Therefore, the shaded area is equal to

$$S_{EFDB} + S_{\triangle ABD} - S_{quarter\ circle} = 2\sqrt{2} - \frac{5}{2} + \frac{1}{2} - \frac{1}{4}\pi = 2\sqrt{2} - 2 - \frac{\pi}{4}.$$

Problem 2

Two parallel chords on the same side of the center of a circle are 12 inches and 20 inches long and 2 inches apart. Find the radius of the circle. Express your answer in simplest radical form.

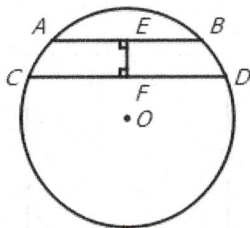

> **Tips:** *1. In problems that involve chords of a circle, it is usually a good idea to construct a right triangle. 2. Connect OA, OC, and OF to form two right triangles.*

Solution 1:

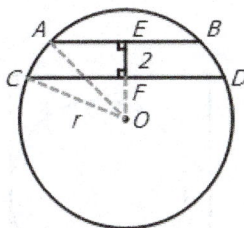

This type of problem can be solved by constructing right triangle and use the **Pythagorean Theorem**.

Connect OA, OC and OF and let $x = OF$. From the triangles $\triangle OAE$ and $\triangle OCF$, we have the equations $r^2 = 6^2 + (2 + x)^2$ and $r^2 = 10^2 + x^2$, respectively.

Solving the equation set, we get $x = 15$ and $r = 5\sqrt{13}$.

Solution 2:

If we focus on the distance between AB and CD, we have $EF = OE - OF = 2$.

Similar to solution 1, we have $OE = \sqrt{r^2 - 6^2}$ and $OF = \sqrt{r^2 - 10^2}$. Thus
$$\sqrt{r^2 - 6^2} - \sqrt{r^2 - 10^2} = 2.$$

It might require some effort to solve this equation, but there are multiple ways to solve it.

If we notice $(\sqrt{r^2 - 6^2} - \sqrt{r^2 - 10^2})(\sqrt{r^2 - 6^2} + \sqrt{r^2 - 10^2}) = 10^2 - 6^2$, we have $\sqrt{r^2 - 6^2} + \sqrt{r^2 - 10^2} = 32$.

Therefore, $2\sqrt{r^2 - 6^2} = 34$ and $r = 5\sqrt{13}$.

43

Problem 3

Square $SQUA$ with midpoints M and N of sides SQ and SA, respectively, has an area of 64 square units. What is the number of square units, rounded to the nearest integer, in the area of the largest circle which can be drawn in pentagon $MNAUQ$? (1996 Mathounts National Target)

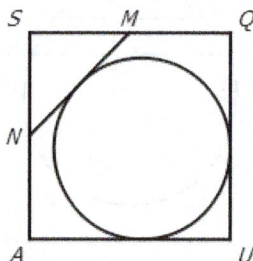

> **Tips:** 1. Connect SU and examine the isosceles right triangles. 2. Connect the center of the circle to the tangent points.

Solution 1:

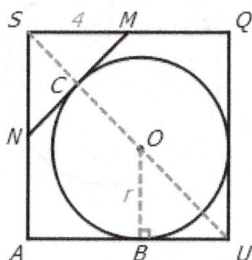

Connect OB and SU.

The area of the square $SQUA$ is 64, which means that the side length is 8. Thus $SM = SN = NA = 4$, $SU = 8\sqrt{2}$, and $SC = 2\sqrt{2}$.

In the isosceles right triangle ΔOBU, $OB = r$ and $OU = \sqrt{2}r$. Therefore, $SU = SC + OC + OU$, which gives the equation $8\sqrt{2} = 2\sqrt{2} + r + \sqrt{2}r$.

Solving for r, we get $r = 12 - 6\sqrt{2}$, so the area of circle is equal to $\pi r^2 \approx 39$.

Solution 2:

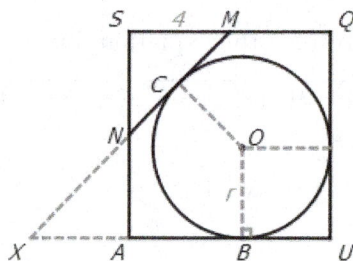

44

Extend MN and UA to meet at point X, then connect XA and XN to use the information from the tangent lines.

Similar to the process in solution 1, we have $SM = SN = NA = 4$ and $NC = 2\sqrt{2}$.

It is easy to notice that $\triangle OBU$ is an isosceles right triangle, so $XN = NM = 4\sqrt{2}$.

From the properties of tangent lines, we know that $XC = $XB. Thus $6\sqrt{2} = 4 + 8 - r$, and $r = 12 - 6\sqrt{2}$.

Therefore, the area of circle is equal to $\pi r^2 \approx 39$.

Problem 4

In square $ABCD$, points E and F are on sides CD and CB, respectively. The angle $\angle EAF = 45°$. If point H is on EF and $AH \perp EF$, prove $AH = AB$.

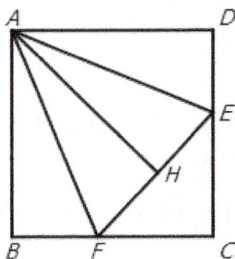

Tips: *1. Think about how you can use the 45 degree information without using trigonometry. 2. Rotate triangle AEF around A by 90 degree clockwise.*

Solution 1:

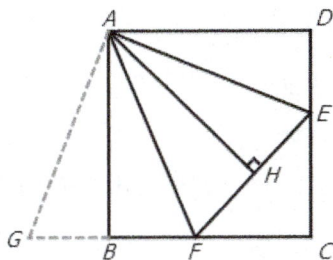

Rotate $\triangle ADE$ $90°$ clockwise around point A. Because $ABCD$ is a square, point D will move to point B and point E will end up at point G.

$\because \triangle ADE \cong \triangle ABG$

$\therefore AG = AE$ and $\angle DAE = \angle BAG$

$\because \angle EAF = 45°$ and $\angle DAB = 90°$

$\therefore \angle DAE + \angle FAB = 45°$

45

$$\therefore \angle FAG = \angle BAG + \angle FAB = \angle DAE + \angle FAB = 45° = \angle EAF$$

$$\because AG = AE, \angle FAG = \angle EAF, \text{ and } \Delta EAF \text{ and } \Delta GAF \text{ share edge } AF$$

$$\therefore \Delta EAF \cong \Delta GAF$$

$$\therefore \angle EFA = \angle GFA$$

$$\because \angle AHF = \angle ABF = 90° \text{ and } \Delta HAF \text{ and } \Delta BAF \text{ share edge } AF$$

$$\therefore \Delta HAF \cong \Delta BAF$$

$$\therefore AH = AB$$

Solution 2:

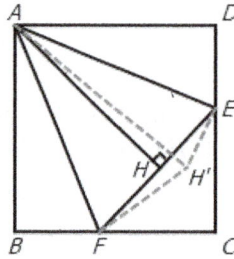

Looking at the conclusion $AH = AB$, we should try to prove $\Delta AHF \cong \Delta ABF$, starting by constructing two congruent triangles.

Flip ΔABF along its edge AF to the other side as $\Delta AH'F$, then connect $H'E$.

$$\because \Delta ABF \cong \Delta AH'F$$

$$\therefore AH' = AB, \angle BAF = \angle H'AF, \text{ and } \angle FH'A = \angle FBA = 90°$$

$$\because \angle EAF = 45° \text{ and } \angle DAB = 90°$$

$$\therefore \angle DAE + \angle BAF = 45°$$

$$\therefore \angle DAE = 45° - \angle BAF = 45° - \angle H'AF = \angle EAF - \angle H'AF = \angle H'AE$$

$$\because AH' = AB = AD \text{ and } \Delta H'AE \text{ and } \Delta DAE \text{ share edge } AE$$

$$\therefore \Delta H'AE \cong \Delta DAE$$

$$\therefore \angle EH'A = \angle EDA = 90°$$

$$\because \angle EH'A = 90° \text{ and } \angle FH'A = 90°$$

$$\therefore \text{Points } F, H' \text{ and } E \text{ are collinear}$$

Therefore, H' is also on line EF. $AH \perp EF$ and $AH' \perp EF$, so H' and H are the same point. Thus $AH = AH' = AB$.

Solution 3:

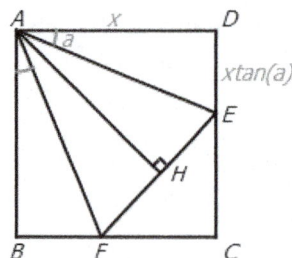

46

Because E is an arbitrary point on CD, trigonometry can be used to calculate the length of AH. Although this may not the simplest solution, the method of trigonometry is a powerful tool that can be used to solve many similar problems.

Let side length of the square be x and the angle $\angle DAE = \alpha$. Then, $DE = x \tan \alpha$, $CE = x - x \tan \alpha$. $BF = x \tan 45° - \alpha$, $CF = x - x \tan 45° - \alpha$. Using the **Pythagorean Theorem**,

$EF^2 = CF^2 + CE^2$

$= x^2[1 - \tan \alpha]^2 + x^2[1 - \tan(45° - \alpha)]^2$

$= x^2[2 - 2\tan\alpha - 2\tan(45° - \alpha) + (\tan\alpha)^2 + (\tan(45° - \alpha))^2]$

$= x^2[2\tan\alpha\tan(45° - \alpha) + (\tan\alpha)^2 + \tan(45° - \alpha)^2]$

$= x^2[\tan\alpha + \tan(45° - \alpha)]^2$

So $EF = x[\tan\alpha + \tan(45° - \alpha)]$

In the above derivation, the sum formula for tangents was used: $\tan\alpha + \tan(45 - \alpha) = \tan 45°[1 - \tan\alpha\tan(45° - \alpha)]$, where $\tan 45° = 1$.

The next step is to get the area of $\triangle AEF$.

$S_{\triangle AEF} = S_{\square ABCD} - S_{\triangle ADE} - S_{\triangle ABF} - S_{\triangle CEF}$

$= x^2 - (1/2)x^2\tan\alpha - (1/2)x^2\tan(45° - \alpha) - (1/2)x^2[1 - \tan\alpha][1 - \tan(45° - \alpha)]$

$= x^2 - (1/2)x^2[1 - \tan\alpha\ tan(45° - \alpha)] - (1/2)x^2[1 - \tan\alpha][1 - tan(45° - \alpha)]$

$= (1/2)x^2[\tan\alpha + \tan(45° - \alpha)]$

Therefore, the altitude of $\triangle AEF$, $AH = 2S_{\triangle AEF}/EF = x = AB$.

Problem 5

In squares $ABCD$ and $BEFG$, point E is on AB and $AD = 2$. Find the area of $\triangle AFC$.

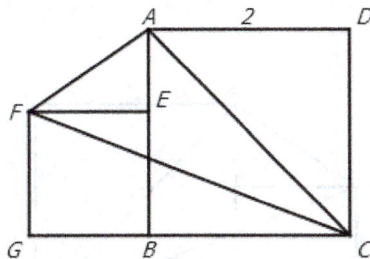

> **Tips:** *1. Connect BF and find parallel lines. 2. Find the area when BEFG is infinitely small.*

Solution 1:

47

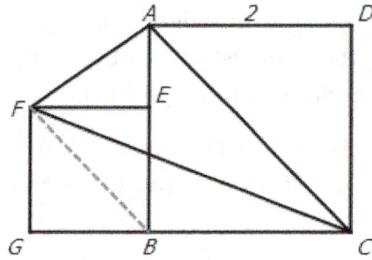

First, connect FB. Both FB and AC are diagonal lines of squares $BEFG$ and $ABCD$, respectively, and both $\triangle ABC$ and $\triangle FGB$ are isosceles right triangles.

$\angle FBG = \angle ACB = 45° \Rightarrow FB // AC$

Thus area of $\triangle AFC$ = area of $\triangle ABC = (1/2) * 2^2 = 2$.

Solution 2:

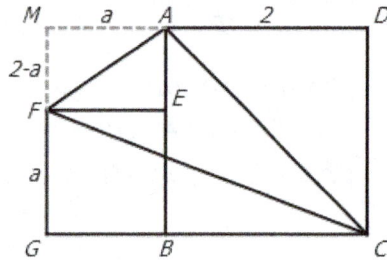

Extend DA and GF to meet at point M to construct rectangle $DCGM$.

Assuming the length of the edge of square $EBGF$ is a, we have

$S_{\triangle AFC} = S_{DCGM} - S_{FGC} - S_{ADC} - S_{AMF} = 2(a+2) - (1/2)a(a+1) - (1/2)2^2 - (1/2)a(2-a) = 2$.

Solution 3:

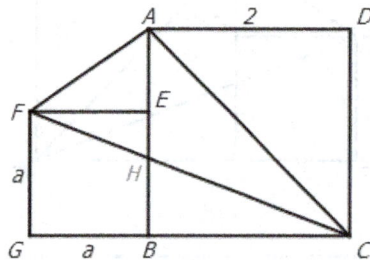

$\triangle AFC$ can be split into two triangles: $\triangle ACH$ and $\triangle AFH$. Since $AB//FG$, $\triangle CHB \sim \triangle CFG$ which leads to

$$HB = \frac{BC}{GC} * FG = \frac{2}{2+a}$$

48

$$AH = 2 - HB = 2 - \frac{2a}{2+a} = \frac{4}{2+a}$$

Thus

$$S_{\triangle AFC} = S_{\triangle ACH} + S_{\triangle AFH}$$

$$= \frac{1}{2} * AH * FE + \frac{1}{2} * AH * BC$$

$$= \frac{2a}{2+a} + \frac{4}{2+a} = 2$$

Solution 4:

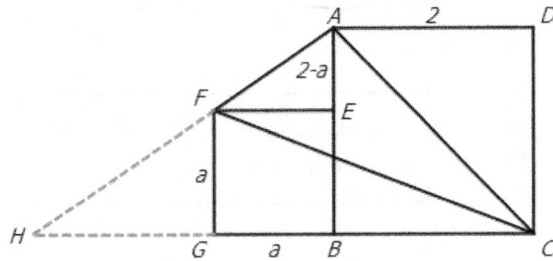

Extend AF and BG to meet at point H. It is easy to show that $\triangle AEF \sim \triangle ABH$. Let the edge length of the square $EBGF$ be a, then

$$\frac{AF}{AH} = \frac{AE}{AB} = \frac{2-a}{2}$$

$$\frac{BH}{EF} = \frac{AB}{AE} = \frac{2}{2-a}$$

Thus

$$BH = \frac{2a}{2-a}$$

$$CH = BH + BC = 2 + \frac{2a}{2-a} = \frac{4}{2-a}$$

Because $\triangle AFC$ and $\triangle AHC$ share the same height, the ratio of their areas is the ratio of their bases:

$$\frac{S_{\triangle AFC}}{S_{\triangle AHC}} = \frac{AF}{AH} = \frac{2-a}{2}$$

Therefore,

$$S_{\triangle AFC} = \frac{2-a}{2} * S_{\triangle AHC} = \frac{(2-a)}{2} * \frac{1}{2} * CH * AB = \frac{(2-a)}{2} * \frac{1}{2} * \frac{4}{(2-a)} * 2 = 2$$

Problem 6

AD is a median of $\triangle ABC$ with D on edge BC. E is on AD such that $AE : AD = 1 : 3$. The line CE intersects edge AB at point F. If the length of AF is 1.2 inch, find the length of segment AB.

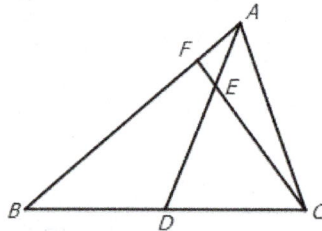

> **Tips:** *1. Create parallel lines when median exists. 2. Use the mass point method. 3. Use the area method to calculate segment ratios.*

Solution 1:

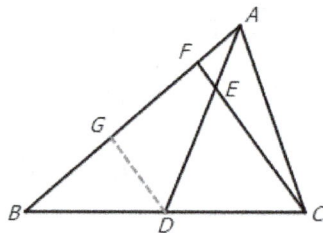

Since D is the median, we can construct a parallel line segment based on midpoint D.

Construct a line DG such that $DG//CF$ and G is on the line AB.

In $\triangle ADG$, $DG//EF$, so $AF/FG = AE/ED = 1 : 2$.

In $\triangle BCF$, $DG//CF$, so $FG/GB = CD/DB = 1 : 1$.

Therefore, $FB = FG + GB = 2FG = 4AF$.

Thus $AB = AF + FB = 5AF = 6$ inches.

Solution 2:

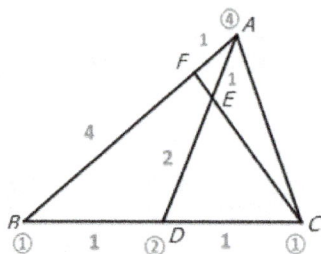

50

This is a perfect example to apply the mass point method.

Let the mass at point B be 1.

Because AD is the median, $BC = CD$. Hence, the mass at point C is 1 and the mass at point D is 2.

Because $AE/AD = 1 : 3$, $AE/ED = 1 : 2$. Thus the mass at point A is 4.

The ratio of the mass at A to the mass at B is 4:1, hence $FB = 4AF$.

Therefore, $AB = AF + FB = 5AF = 6$ inches

Solution 3:

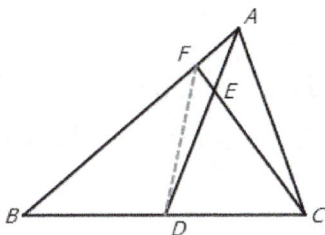

This problem can also be solved by the area method. Connect D and F. Examine the area of four triangles: $\triangle AFC$, $\triangle DFC$, $\triangle BDF$, and $\triangle BCF$.

From the area method, we know

$S_{\triangle DFC} : S_{\triangle AFC} = DE : EA = 2 : 1$

$S_{\triangle BDF} : S_{\triangle DFC} = BD : DC = 1 : 1$

Thus $S_{\triangle BCF} : S_{\triangle AFC} = 4 : 1$

$BF : FA = S_{\triangle BCF} : S_{\triangle AFC} = 4 : 1$

Therefore, $AB = 5FA = 6$ inches.

Problem 7

AM is the median of $\triangle ABC$ and M is on side BC. Points P and Q are on edge AB and AC, respectively. Given $AC : QC = 4 : 1$ and $AB : AP = 5 : 2$, find $AM : AN$.

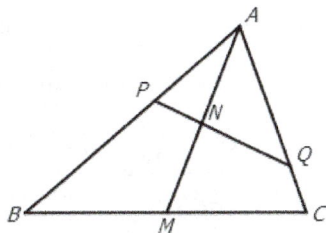

Solution 1:

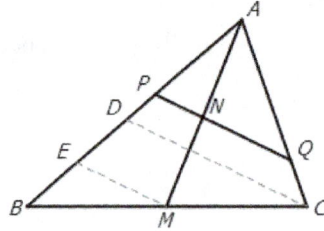

Create multiple parallel lines to transfer ratio information to edge AB.

Construct lines CD and ME such that $CD//ME//PQ$, and point D and E are on the line AB.

In $\triangle ACD$, $CD//PQ$. Thus $PD/AP = CQ/QA = 1:3$

In $\triangle BCD$, $ME//CD$. Thus $DE/EB = CM/MB = 1:1$

The condition $AB/AP = 5:2$ leads to $BP/AP = 3:2$

Thus,

$$PE = PD + DE = \frac{1}{3}AP + \frac{1}{2}(\frac{3}{2} - \frac{1}{3})AP = \frac{11}{12}AP$$

In $\triangle AEM$, $ME//PQ$. Thus $AN/NM = AP/PE = 12/11$

Therefore, $AM/AN = (1 + 12/11)/(12/11) = 23/12$

Solution 2:

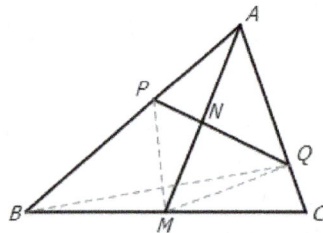

Use the area method. The ratio of the bases equal to the ratio of triangle areas.

Connect PM, QM, and BP to calculate area of multiple triangles.

$$\frac{S_{\triangle ABQ}}{S_{\triangle ABC}} = \frac{AQ}{AC} = \frac{3}{4}$$

$$\frac{S_{\triangle APQ}}{S_{\triangle ABP}} = \frac{AP}{AB} = \frac{2}{5}$$

Thus

$$\frac{S_{\triangle APQ}}{S_{\triangle ABP}} = \frac{3}{4} * \frac{2}{5} = \frac{3}{10}$$

Similarly, we can get

$$\frac{S_{\triangle PBM}}{S_{\triangle ABC}} = \frac{1}{2} * \frac{3}{5} = \frac{3}{10}$$

$$\frac{S_{\triangle QCM}}{S_{\triangle ABC}} = \frac{1}{2} * \frac{1}{4} = \frac{1}{8}$$

Thus

$$\frac{S_{\triangle PQM}}{S_{\triangle ABC}} = 1 - \frac{3}{10} - \frac{1}{5} - \frac{1}{8} = \frac{11}{40}$$

$$\frac{AN}{NM} = \frac{S_{\triangle APQ}}{S_{\triangle PQM}} = \frac{12}{11}$$

Therefore,

$$AM/AN = 23/12$$

Solution 3:

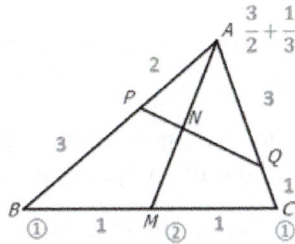

Use the mass point method, and let the mass at point B and C be 2. The mass at point M is 2 and the mass at point A is $\frac{3}{2} + \frac{1}{3} = \frac{11}{6}$.

Thus $AN/MN = 12/11$ and $AM/AN = 23/12$ (the mass point method can be very powerful).

Problem 8

In the diagram below find $\angle A + \angle B + \angle C + \angle D + \angle E + \angle F + \angle G$.

(A) 360°   (B) 450°   (C) 540°   (D) 720°

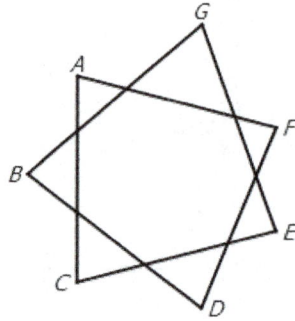

Tips: *Create a polygon and use the properties of exterior angles for a polygon.*

Solution:

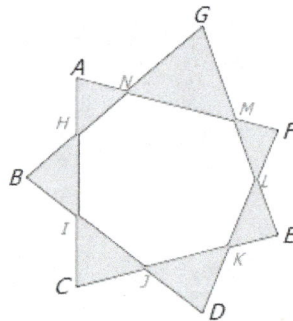

The angles of interest belong to the 7 triangles shaded green. Except the angles at $A, B, ..., G$, all the other angles are exterior angles of the polygon $HIJKLMN$.

The sum of the degrees of these exterior angles is $2 \times 360°$ (each exterior angle appears twice). Therefore, $\angle A + \angle B + \angle C + \angle D + \angle E + \angle F + \angle G = 7 * 180° - 2 * 360° = 540°$

The answer is (C).

Problem 9

CD is the altitude of $Rt\triangle ABC$ on edge AB. AE is the angular bisector of $\angle CAB$ and intersect edge BC at point E and CF is the angular bisector of $\angle BCD$ and intersect edge DB at point F. Show $HF//BC$.

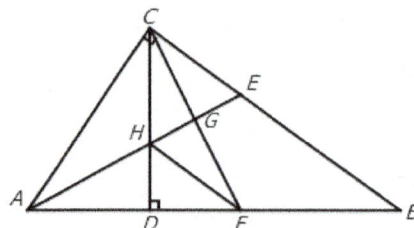

Tips: *1. To prove two parallel lines, we only need to prove the equal ratio of segments or equal angles. 2. Use ratio properties from angular bisector.*

Solution 1:

To prove $HF//BC$, it is sufficient to prove $DH/CH = DF/FB$.

$\because CF$ is the angular bisector of $\angle BCD$

$\therefore DF/FB = CD/CB$, and

$\because AE$ is the angular bisector of $\angle CAD$

$\therefore DH/HC = AD/AC$

In $Rt\triangle ABC$, CD is the altitude which leads to $\triangle BCD \sim \triangle CAD$.

$\therefore CD/CB = AD/AC$

$\therefore DF/FB = DH/HC$

$\therefore HF//BC$

Solution 2:

It is also sufficient to prove $\angle CFH = \angle BCF$ to show $HF//BC$.

In $Rt\triangle ABC$, CD is the altitude which leads to $\triangle BCD \sim \triangle CAD$. Thus $\angle BCD = \angle CAD$. CF and AE are the angular bisectors of $\angle BCD$ and $\angle CAD$, respectively.

Therefore, $\angle BCF = \angle DCF = \angle CAE = \angle BAE$.

In $Rt\triangle ACD$, $\angle CAD + \angle DCA = 90°$ and $\angle DCF = \angle CAE$, thus $\angle CAE + \angle ACF = 90°$

$\therefore AE \perp CF$

$\because AE$ is the angular bisector of $\angle CAD$

$\therefore \triangle CAF$ is an isosceles triangle

$\therefore AG$ perpendicular bisector of edge CF

$\therefore \triangle CHF$ is also an isosceles triangle

$\therefore \angle HFC = \angle HCF = \angle FCB$

$\therefore HF//BC$

Problem 10

In the diagram below $\angle BGF = \alpha$, find $\angle A + \angle B + \angle C + \angle D + \angle E + \angle F$.

(A) $360° - \alpha$ (B) $270° - \alpha$ (C) $180° - \alpha$ (D) 2α

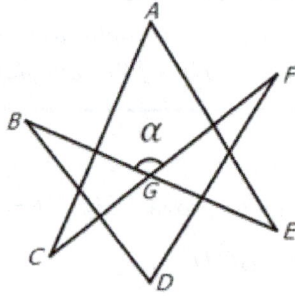

Tips: *1. Divide all the angles into two groups. 2. Use the properties of the exterior angles for a triangle.*

Solution:

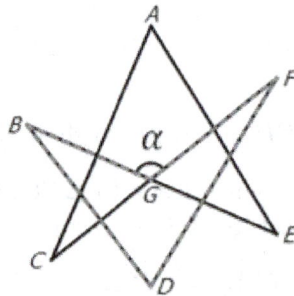

For the quadrilateral $BDFG$, α is the exterior angle and $\alpha = \angle B + \angle D + \angle F$.

Similarly, $\angle CGE = \angle A + \angle C + \angle E$.

We also know that $\angle CGE = \alpha$. Therefore, $\angle A + \angle B + \angle C + \angle D + \angle E + \angle F = 2\alpha$.

The answer is (D).

Problem 11

In the $\triangle ABC$, $\angle BAC = 90° + B$, the angular bisector of $\angle ACB$ intersects the edge AB at point L and the exterior angular bisector CN intersects the edge AB at point N. If the length of $CL = 3$ cm, what is the length of segment CN?

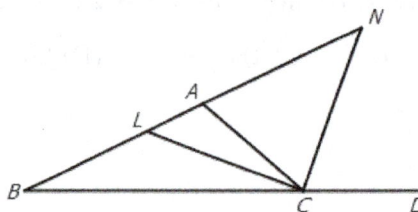

56

Tips: *1. Prove triangle LCN is an isosceles right triangle.*

Solution 1:

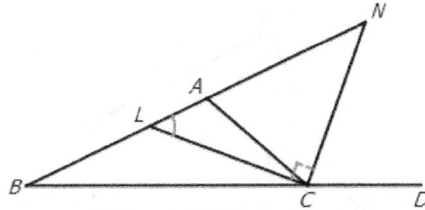

In the $\triangle ABC$, CL and CN are interior and exterior angular bisectors of $\angle ACB$. Thus $\angle LCN = 90°$. CL is the angular bisector of $\angle ACB$, thus $\angle BCL = \angle ACL$.

$$\angle ALC = \angle B + \angle BCL = \angle B + \angle BCA = \frac{1}{2}\angle B + \frac{1}{2}(\angle B + \angle BCA)$$

$$= \frac{1}{2}\angle B + \frac{1}{2}(180° - \angle BAC) = \frac{1}{2}\angle B + \frac{1}{2}[(180° - (90° + \angle B)] = 45°$$

Therefore, $\triangle LCN$ is an isosceles right triangle, and $CN = CL = 3$ cm.

Solution 2:

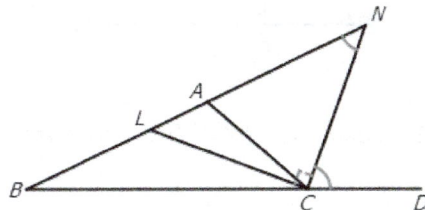

Find angle N first.

In the $\triangle ABC$, CL and CN are interior and exterior angular bisector of $\angle ACB$. Thus $\angle LCN = 90°$.

Because CN is the angular bisector of $\angle ACD$, $\angle DCN = \frac{1}{2}\angle ACD$.

$\angle ACD$ is exterior angle, which leads to $ACD = \angle CAB + \angle B = 90° + 2\angle B$.

Thus $\angle DCN = \frac{1}{2}\angle ACD = 45° + \angle B$ and $\angle N = \angle DCN - \angle B = 45°$

Therefore, $\triangle LCN$ is an isosceles right triangle, and $CN = CL = 3$ cm.

Problem 12

In quadrilateral $ABCD$, $AB = AD$, $AB \perp AD$, $CD \perp BD$ and $AE \perp BC$ with E on edge BC. If the length of AE is 1, find the area of $ABCD$.

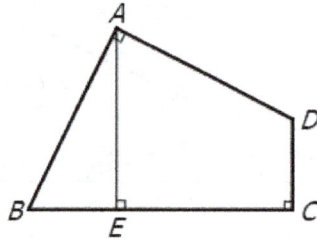

Tips: 1. Think about how to use 45 degree information without using trigonometry. 2. Rotate triangle ABE to form a square.

Solution 1:

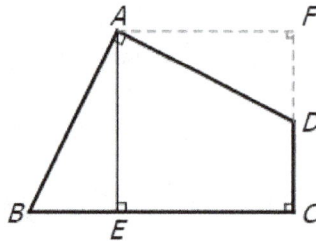

Construct a parallel line AF such that $AF//BC$ and intersect the extension of edge CD at point F.

Because $AF//BC$, $CD \perp BD$ and $AE \perp BC$, $AECF$ is a rectangle and $\angle F = \angle FAE = 90°$.

With the condition $AD \perp AB$, we get $\angle FAD = \angle EAB$. We also know $AB = AD$, thus $\triangle AFD \cong \triangle AEB$, which leads to $AF = AE = 1$ and $S_{\triangle AFD} = S_{\triangle AEB}$.

Therefore, $AECF$ is a square with edge-length. The area of $ABCD$ is equal to the area of square $AECF$, which is 1.

Solution 2:

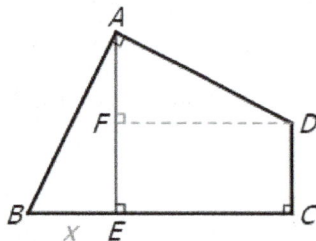

58

Cut $ABCD$ into pieces and calculate the area of each individual piece.

As shown above, construct a line DF such that $DF//AE$ with F on line AE. It is easy to see $CDEF$ is a rectangle.

Because $AD \perp AB$, $AE \perp BC$ and $AB = AD$, we have $\triangle DFA \cong \triangle AEB$ and $DF = AE = CE = 1$.

Let the length of BE be x. The area of $ABCD =$

$$S_{\triangle AFD} + S_{\triangle AEB} + S_{\square CDFE} = \frac{1}{2} * x * 1 + \frac{1}{2} * x * 1 + 1 * (1 - x) = 1$$

Problem 13

Given that AB is tangent to circle P at B. what is the ratio $\frac{x}{x+y}$? Express your answer as a common fraction. (2000 Mathcounts State Sprint).

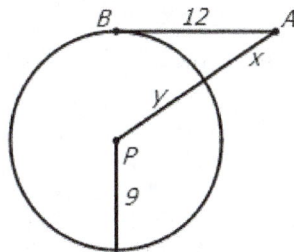

> **Tips:** *1. Extend AP and use the Power of a Point Theorem. 2. Connect PB and use the Pythagorean theorem.*

Solution 1:

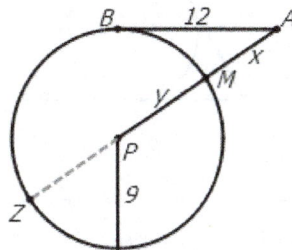

The tangent line segment AB indicates that we can use the **Power of a Point Theorem** to solve this problem. Extend AP to meet the circle at point Z. AB and AZ are tangent and secant lines to circle P.

The **Power of a Point Theorem** gives $AB^2 = AM * AZ$. Thus $12^2 = x * (x + 9 + 9)$, which leads to $x = 6$ and

59

$$\frac{x}{x+y} = \frac{6}{6+9} = \frac{2}{5}$$

Solution 2:

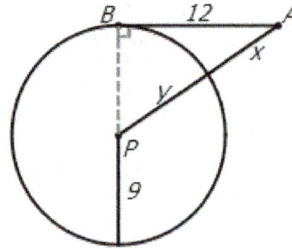

Connect the center of circle P and the tangent point B to construct a right triangle $\triangle ABP$. In the $\triangle ABP$, the **Pythagorean Theorem** gives us $(x+9)^2 = 9^2+12^2$. Solving the equation gives $x = 6$ and

$$\frac{x}{x+y} = \frac{6}{6+9} = \frac{2}{5}$$

You may notice that $\triangle ABP$ is a multiple of a 3, 4, 5 triangle, and obtain AP = 15 directly.

Problem 14

$\triangle ABC$ is an equilateral triangle where O is the center of its inscribed circle. If the area of the circle is 4π cm^2, what is the area, in square centimeters, of the triangle ABC? Express your answer in simplest radical form. (1992 Mathcounts State Sprint).

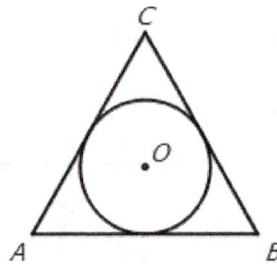

Tips: *1. Connect the center of the circle to a vertex of the triangle and to the tangential point to create a right triangle. 2. The incenter of a triangle is at the intersection of the angle bisectors.*

Solution 1:

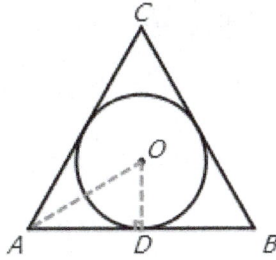

Since an equilateral triangle has only one independent variable (edge length), we just need to find the edge length.

The circle's area $S = \pi r^2 = 4\pi$, so $r = 2$ cm.

Connect OA and create a $Rt\triangle OAD$ as shown in the diagram. $\angle OAD = 30°$, $OD = 2$ cm.

Therefore, $AD = 2\sqrt{3}$ and $AB = 4\sqrt{3}$.

The area of the equilateral triangle is

$$\frac{\sqrt{3}}{4} * \left(4\sqrt{3}\right)^2 = 12\sqrt{3}$$

Solution 2:

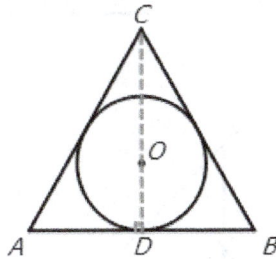

We can find the altitude CD easily. In an equilateral triangle, the centers (incenter, centroid, circumcenter and orthocenter) are all located at the same point.

Because O is the centroid, $CD = 3OD$. The circle's area $S = \pi r^2 = 4\pi$, so $CD = 6$ cm.

$\angle ACD = 30°$, so $AD = \frac{\sqrt{3}}{3}CD = 2\sqrt{3}$ and the area of equilateral triangle is $12\sqrt{3}$.

Solution 3:

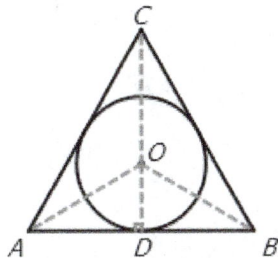

From solution 2, we know $OD = 2$ and $AB = 4\sqrt{3}$.

Thus $S_{\triangle ADE} = \frac{1}{2}(AB + BC + CA) * r = 12\sqrt{3}$.

Problem 15

$\triangle ABC$ is an isosceles right triangle with hypotenuse AC. A circle is drawn through B that is tangent to AC at P. PB is a diameter of the circle. Given that AC is 8 units long, how many square units are in the area of the region inside the circle but outside the triangle? Express your answer in terms of π. (1996 Mathcounts National Sprint).

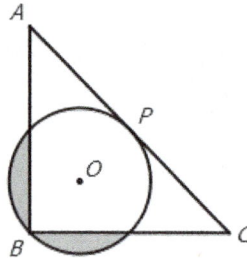

Tips: 1. Connect the center of circle O to the tangent and intersection points.

Solution 1:

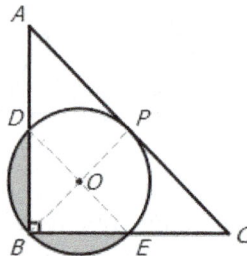

For isosceles $Rt\triangle ABC$, we have $\angle A = \angle C = 45°$.

Because circle O is tangent to AC and BP is the diameter, $BP \perp AC$, which leads to two isosceles $Rt\triangle ABP$ and $Rt\triangle CBP$.

So $BP = AP = CP = 4$ units. With $\angle ABC = 90°$, DE is also a diameter with length of 8 units.

The shaded area $= \frac{1}{2}\pi r^2 - OB * OD = 2\pi - 4$

Solution 2:

62

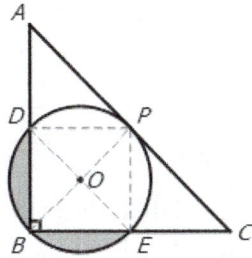

The shaded area is equal to half of the difference between the area of the full circle and the area of inscribed square $BEPD$.

Using a similar method as solution 1, we get circle radius of 2 and square area of $\frac{1}{2} * 4^2$, so the shaded area is $2\pi - 4$.

Problem 16

One side of a triangle is divided into segments of lengths 6 cm and 8 cm by the point of tangency of the inscribed circle. If the radius of the circle is 4 cm, what is the length, in centimeters, of the shortest side of the triangles? (1992 National Team).

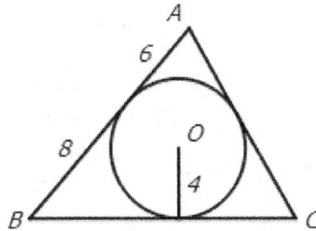

> **Tips:** *1. Apply Heron's Formula. 2. Use trigonometric identities for the tangent function.*

Solution 1:

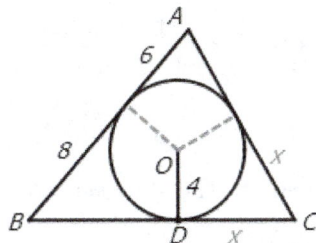

The fundamental constraints in this problem are area conservation (area method) and intrinsic trigonometric identities.

63

The way to use the area method is to calculate the same area by different methods. Assume the length of segment CD is x.

The area calculated using **Heron's Formula** is $A = \sqrt{s(s-a)(s-b)(s-c)}$ where $s = \frac{a+b+c}{2}$. $s = (6+8+x)$ and $A = \sqrt{(14+x)x*6*8}$.

The triangle's area can also be calculated through multiplying the semi perimeter by the radius of the inscribed circle: $A = s*r = 4*(14+x)$.

Thus we have $\sqrt{48*x(x+14)} = 4(x+14)$. Solving the equation by squaring on both sides gives $48x = 16(x+14)$ and $x = 7$.

Therefore, the length of the shortest side of the triangle is $7 + 6 = 13$.

Solution 2:

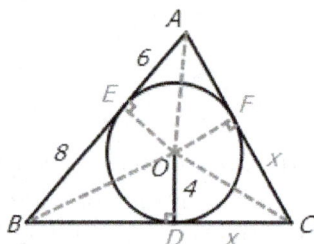

We can also use trigonometry to solve the problem.

For an inscribed circle, we know OA, OB and OC are all angular bisectors of the angles A, B, and C, respectively, so we have $\angle OAE + \angle OBE + \angle OCD = 90°$.

In the right triangles $\triangle OAE$, $\triangle OBE$ and $\triangle ODC$, we have

$$\tan(\angle OAB) = \frac{4}{6} = \frac{2}{3}$$

$$\tan(\angle OBA) = \frac{4}{8} = \frac{1}{2}$$

and

$$\tan(\angle OCD) = \frac{4}{x}$$

Thus

$$\tan(\angle OAE + \angle OBE) = \frac{\tan(\angle OAE) + \tan(\angle OBE)}{1 - \tan(\angle OAE)*\tan(\angle OBE)} = \frac{4}{7}$$

Because $OAE + \angle OBE + \angle OCD = 90°$, $\cot(\angle OCD) = \tan(\angle OAE + \angle OBE)$.

Thus $\frac{4}{x} = \frac{4}{7}$ and $x = 7$, which leads to the shortest side of the triangle being $7 + 6 = 13$.

Problem 17

A circular table is pushed into the corner of a square room so that a point P on the edge of the table is $8''$ from one wall and $9''$ from the other wall as shown. Find the radius of the table in inches. (1994 Mathcounts National Target).

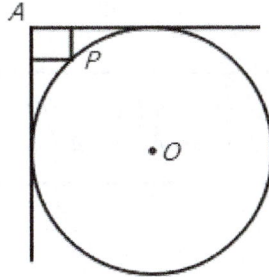

Tips: *1. The walls are tangent to the table. 2. Connect the tangent points to the center of the circle to form right triangles.*

Solution:

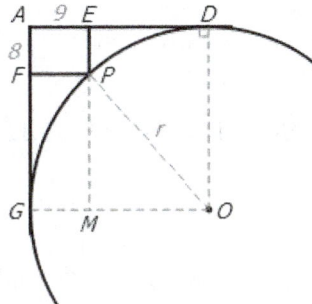

Connect center O to two tangent points D and G and connect OP and PM.

$ADOG$ is a square with edge length of r. In the $Rt\triangle OPM$, $OP = r$, $OM = r - 9$, and $PM = r - 8$. By the **Pythagorean Theorem**, $(r - 8)^2 + (r - 9)^2 = r^2$.

Solving this quadratic equation gives $r = 29$.

Problem 18

In triangle $\triangle ABC$, point D divides side AC so that $AD : DC = 1 : 2$. Let E be the midpoint of BD and let F be the point of intersection of line BC and line AE. Given that the area of $\triangle ABC$ is 360, what is the area of $\triangle EBF$? (2019 AMC8).

(A) 24   (B) 30   (C) 32   (D) 36   (E) 40

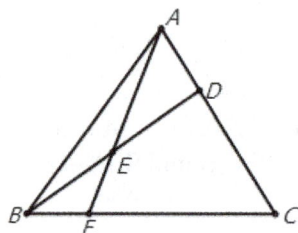

Tips: *1. Construct a line through point D parallel to AF. 2. Use the area method. 3. Use the mass point method to find segment ratios.*

Solution 1:

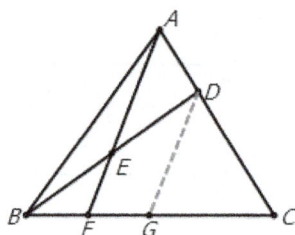

Drawing parallel lines is helpful because it transfers outside information (e.g. $AD : DC = 1 : 2$ here) to inside.

Draw $DG // AF$ and G is on BC. Because $DG // AF$, $CG/FG = CD/AD = 2 : 1$ and $FG/FB = DE/EG = 1/1$.

Therefore, $CG = BG = 2BF$.

It is then easy to find $S_{\triangle BDC} = 240$, $S_{\triangle BDG} = 120$ and $S_{\triangle EBF} = 30$.

The answer is (B).

Solution 2:

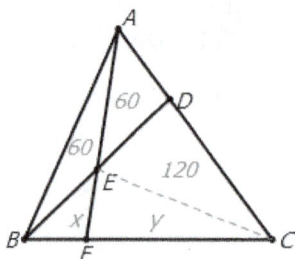

To solve this problem by the area method, it is necessary to first divide the original triangle into smaller triangles by connecting EC.

Because $AD/DC = 1 : 2$ and E is the midpoint of BD, $S_{\triangle EAD} = S_{\triangle EAB} = 60$, $S_{\triangle EDC} = 120$, $S_{\triangle EAC} = 180$, and $S_{\triangle EBF} + S_{\triangle EFC} = 120$, which is $x + y = 120$.

From the area method we also know that

$$\frac{S_{\triangle EAC}}{S_{\triangle EAB}} = \frac{FC}{FB} = \frac{y}{x} = \frac{180}{60}$$

Therefore, $y = 3x$ and $x = 30$.

The answer is (B).

Solution 3:

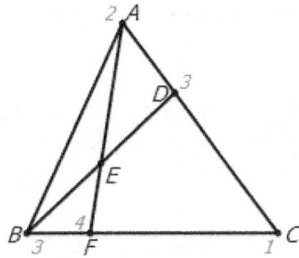

Here, we use the mass point method to calculate CF/BF and AE/EF.

Start by assigning mass of 1 and 2 to point C and A, which gives mass of 3 at point D.

Because E is midpoint of BD, the mass at B and E are 3 and 6, respectively. So mass at point F is 4.

Therefore, $CF/BF = 3 : 1$ and $AE/EF = 2 : 1$. And

$$S_{\triangle EBF} = \frac{1}{3} * S_{\triangle EBF} = \frac{1}{3} * \frac{1}{4} * S_{\triangle ABC} = 360/12 = 30$$

The answer is (B).

Problem 19

In the non-convex quadrilateral $ABCD$ shown below, angle BCD is a right angle, $AB = 12$, $BC = 4$, $CD = 3$, and $AD = 13$. What is the area of quadrilateral $ABCD$? (2017 AMC8)

(A) 12   (B) 24   (C) 26   (D) 30   (E) 36

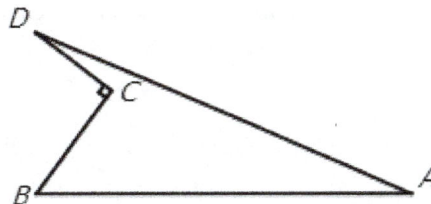

Tips: *Connect DB and find two right triangles.*

Solution:

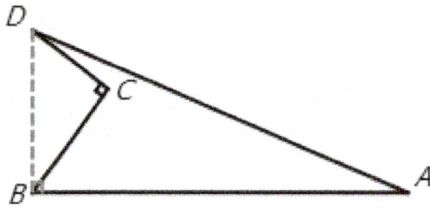

Connect DB to create $Rt\triangle DBC$. Using the **Pythagorean Theorem**, we get $DB = 5$. Along with $AD = 13$ and $AB = 12$, the converse of the Pythagorean Theorem tells us that $\triangle DBC$ is a right triangle.

Therefore, the area of $ABCD$ is $\frac{1}{2} * 5 * 12 - \frac{1}{2} * 3 * 4 = 30 - 6 = 24$.

The answer is (B).

Problem 20

In $\triangle ABC$, a point E is on AB with $AE = 1$ and $EB = 2$. Point D is on AC so that $DE // BC$ and point F is on BC so that $EF // AC$. What is the ratio of the area of $CDEF$ to the area of $\triangle ABC$? (2018 AMC8)

(A) $\frac{4}{9}$ (B) $\frac{1}{2}$ (C) $\frac{5}{9}$ (D) $\frac{3}{5}$ (E) $\frac{2}{3}$

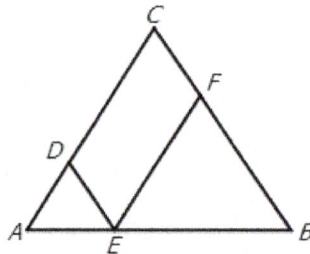

Tips: *1. Use a complementary area method. 2. There are three similar triangles.*

Solution 1:

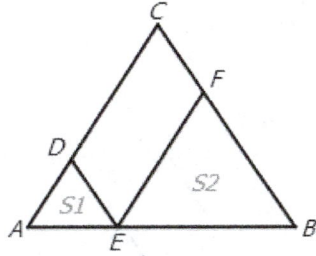

Because $DE//CB$ and $FE//CA$, $\triangle ADE \sim \triangle EFB \sim \triangle ACB$. Let the the area of $\triangle ADE$, $\triangle EFB$ and $\triangle ACB$ be $S1$, $S2$ and $S0$, respectively.

$$\frac{S1}{S0} = \left(\frac{AE}{AB}\right)^2 = \frac{1}{9}$$

and

$$\frac{S2}{S0} = \left(\frac{BE}{AB}\right)^2 = \frac{4}{9}$$

Therefore

$$\frac{S_{CDEF}}{S0} = 1 - \frac{S1}{S0} - \frac{S2}{S0} = 1 - \frac{1}{9} - \frac{4}{9} = \frac{4}{9}$$

The answer is (A).

Solution 2:

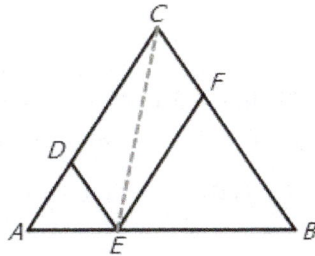

This problem can be solved quickly using the area method.

Connect CE, which is diagonal of parallelogram $CDEF$. Because $DE//CB$, $AD/DC = AE/EB = 1/2$. From the area method, we have

$$\frac{S_{\triangle ACE}}{S_{\triangle ABC}} = \frac{AE}{AB} = \frac{1}{3}$$

and

$$\frac{S_{\triangle CDE}}{S_{\triangle ACE}} = \frac{CD}{CA} = \frac{2}{3}$$

Thus

$$\frac{S_{CDEF}}{S_{\triangle ABC}} = 2 * \frac{S_{\triangle CDE}}{S_{\triangle ABC}} = 2 * \frac{1}{3} * \frac{2}{3} = \frac{4}{9}$$

69

The answer is (A).

Solution 3:

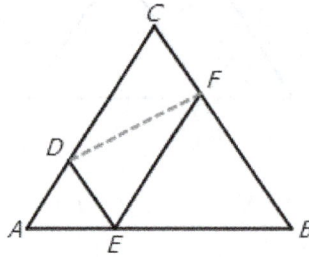

First, connect DF. Comparing the area of $\triangle CDF$ and $\triangle ABC$ which share the angle C, we have

$$\frac{S_{\triangle CDF}}{S_{\triangle ABC}} = \frac{CF * CD}{CB * CA} = \frac{1}{3} * \frac{2}{3} = \frac{2}{9}$$

Therefore,

$$\frac{S_{CDEF}}{S_{\triangle ABC}} = 2 * \frac{S_{\triangle CDF}}{S_{\triangle ABC}} = \frac{4}{9}$$

The answer is (A).

Problem 21

Point E is the midpoint of side CD in square $ABCD$, and BE meets diagonal AC at F. The area of quadrilateral $AFED$ is 45. What is the area of $ABCD$? (2018 AMC8)

(A) 100   (B) 108   (C) 120   (D) 135   (E) 144

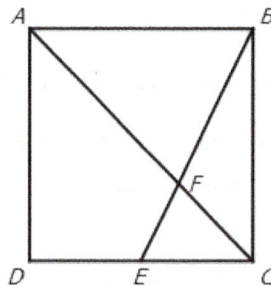

Tips: *The two triangles are similar.*

Solution 1:

70

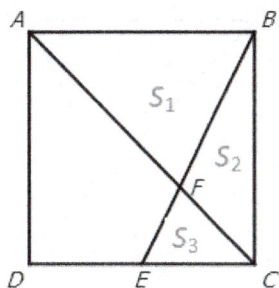

Calculate the ratio between the area of $ADEF$ and the area of square $ABCD$.

Let S_1, S_2, S_3 and S_0 be the area of $\triangle ABF$, $\triangle BCF$ and $\triangle CEF$ and square $ABCD$, respectively. Because E is midpoint of CD and $DC // AB$,

$$\frac{S_3}{S_1} = \frac{1}{4}$$

and

$$\frac{EF}{FB} = \frac{1}{2}$$

Thus $S_1/S_2 = 2$.

The area of $Rt\triangle BCE$ is $(S_2 + S_3) = \frac{1}{4}S_0$, so $S_2 = \frac{1}{6}S_0$ and $S_3 = \frac{1}{12}S_0$. Therefore, $S_1 = \frac{1}{3}S_0$ and $S_{ADEF} = \frac{5}{12}$.

Given the area of $ADEF$ is 45, the area of $ABCD$ is 108.

The answer is (B).

Solution 2:

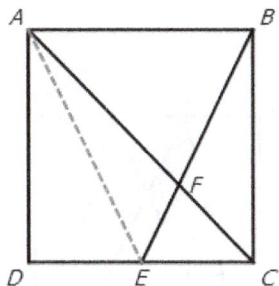

Because $AB // EC$, we have $S_{\triangle AEF} = S_{\triangle BCF}$ and $S_{\triangle ADE} = \frac{1}{4}S_0$.

Similar to solution 1, we have $S_{\triangle BCF} = \frac{1}{6}S_0$. Therefore $S_{\triangle ADEF} = \frac{5}{12}S_0$.

Given the area of $ADEF$ is 45, the area of $ABCD$ is 108.

The answer is (B).

Solution 3:

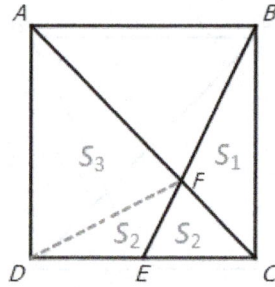

Connect DF. Since E is the midpoint of DC, $S_{\triangle FDE} = S_{\triangle FCE} = S_2$.

$\because AB//EC$ and $AB = 2EC$

$\therefore BF = 2FE$, $AF = 2FC$ and $S_1 = 2S_2$

$\because S_1 + S_2 = \frac{1}{4}S_{ABCD}$

$\therefore S_2 = \frac{1}{12}S_{ABCD}$

Because $AF = 2FC$,

$S_3 = 2(S_2 + S_2) = \frac{1}{3}S_{ABCD}$ and $S_{ADEF} = S_3 + S_2 = (\frac{1}{3} + \frac{1}{12})S_{ABCD}$.

Given the area of $ADEF$ of 45, the area of $ABCD$ is 108.

The answer is (B).

Problem 22

Rectangle $DEFA$ below is a 3×4 rectangle with $DC = CB = BA$. What is the area of the "bat wings" (shaded area)? (2016 AMC8)

(A) 2   (B) 2.5   (C) 3   (D) 3.5   (E) 4

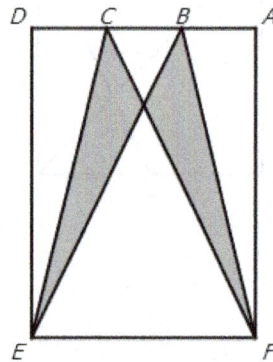

> **Tips:** *Find similar triangles in the diagram and get segment ratios.*

Solution 1:

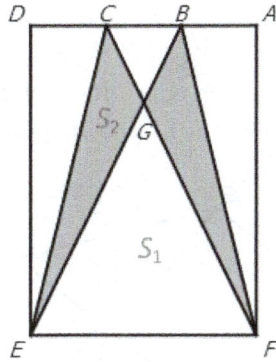

In the square $ABCD$, we have $AB = BC = CD = 1$, $AF = ED = 4$ and $AD//EF$.

$\because AD//EF$ and $EF = 3BC$

$\therefore GF = 3GC$ and $S_1 = 3S_2$

The area of $\triangle CEF$ is equal to $\frac{1}{2} * 3 * 4$ =6. So $S_2 = \frac{1}{4}S_{\triangle CEF} = \frac{3}{2}$.

Therefore, the area of the "bat wings" is 3.

The answer is (C).

Solution 2:

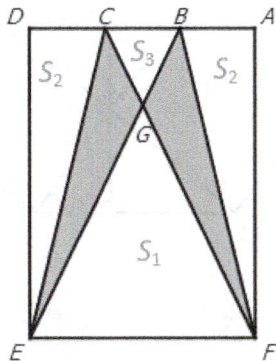

The area of the "bat wings" is equal to the area of the rectangle minus the area of the 4 triangles.

In the diagram above, it is easily seen that $S_2 = 2$.

Because $AD//EF$ and $EF = 3BC$, $\triangle CBG \sim \triangle FEG$ and $S_1 = 9S_3$.

The altitude of base EF is $\frac{3}{4} * AF = 3$ and $S_1 = 4.5$ and $S_2 = 0.5$.

Therefore, the area of "bat wings" is $12 - 2 - 2 - 4.5 - 0.5 = 3$.

The answer is (C).

Problem 23

Square $ACGF$ and $BCDE$ are on the edge AC and BC of $\triangle ABC$, respectively. CH is the altitude of base AB and intersect segment DG at point P. Prove $DP = GP$.

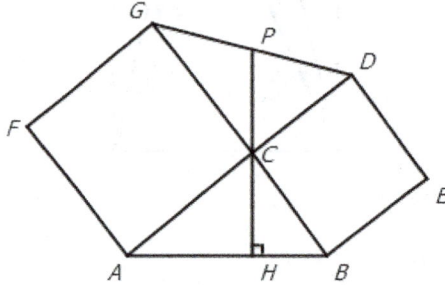

Tips: *1. This is a very challenging problem where you need to construct multiple congruent right triangles. 2. Start from the conclusion by constructing congruent triangles around the congruent segments, DP and GP.*

Solution 1:

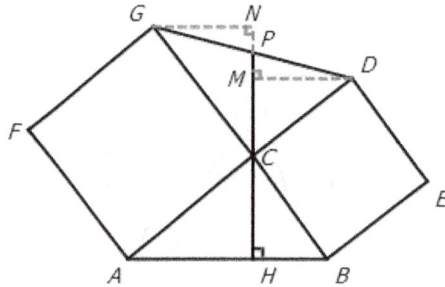

To prove $DP = GP$, construct triangles with DP and GP as an edge and prove these two triangles are congruent.

Construct DM and GN such that $DM \perp PC$, $GN \perp PC$, and M and N are on line PC.

$\because BCED$ is a square

$\therefore CD = BC$ and $\angle DCB = 90°$

$\therefore \angle DCM + \angle BCH = 90°$

$\because CN \perp AB$

$\therefore \angle CBH + \angle BCH = 90°$

$\therefore \angle DCM = \angle CBH$

In $\triangle BCH$ and $\triangle CDM$, we have $BC = CD$, $\angle DCM = \angle CBH$ and $\angle DMC = \angle BHC$.

$\therefore \triangle BCH \cong \triangle CDM$

$\therefore DM = CH.$

74

Using the same process as above, we can prove $GN = CH$.

$\therefore DM = GN$

In $\triangle DMP$ and $\triangle GNP$, we have $DM = GN$, $\angle DPM = \angle GPN$ and $\angle DMC = \angle GNP$.

Thus $\triangle DMP \cong \triangle GNP$ and $DP = GP$

Solution 2:

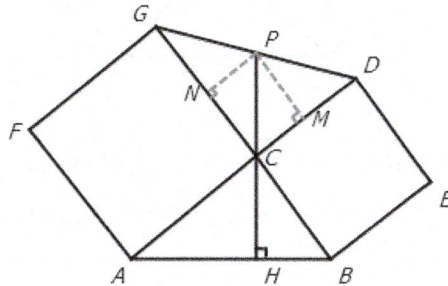

This problem can also be solved efficiently using the area method. To prove $DP = GP$, you just need to prove that the area of $\triangle CDP$ and $\triangle CGP$ are equal.

Construct PM and PN such that $PM \perp CD$ and $PN \perp CG$.

$\because BCED$ is a square

$\therefore CD = BC$ and $\angle DCB = 90°$

$\therefore \angle DCM + \angle BCH = 90°$

$\because CN \perp AB$

$\therefore \angle CBH + \angle BCH = 90°$

$\therefore \angle DCM = \angle CBH$

In $\triangle BCH$ and $\triangle CPM$, we have $\angle DCM = \angle CBH$ and $\angle DMC = \angle BHC$.

$\therefore \triangle BCH \sim \triangle CDM$

$\therefore \frac{PM}{CH} = \frac{CP}{BC}$

$\therefore PM * BC = CP * CH$

$\therefore PM * CD = CP * CH$

Using the same process, we can prove $PNCG = CPCH$.

$\therefore PM * CD = PN * CG$

Therefore, the area of $\triangle CDP$ and $\triangle CGP$ are equal, so $DP = GP$

Solution 3:

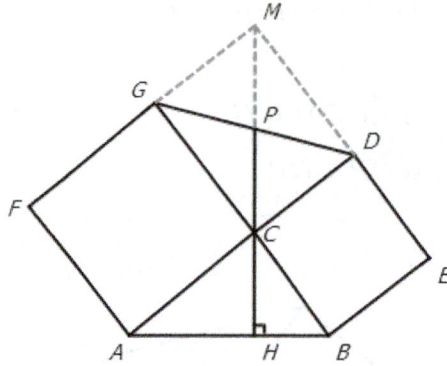

Construct a parallelogram and make CP half of its diagonal.

As shown in the diagram above, make $DM//CG$ and $GM//CD$ and intersect at point M. Connect MP.

$BCDE$ and $ACGF$ are squares, so $\angle BCD = 90°$ and $\angle ACG = 90°$ and $\angle BCA + \angle DCG = 180°$.

$\because DM//CG$ and $GM//CD$

$\therefore \angle CDM + \angle DCG = 180°$ and $\angle CGM + \angle DCG = 180°$

$\therefore \angle BCA = \angle CDM = \angle CGM$, $MD = CG$, $CD = GM$

$\therefore \triangle ABC \cong \triangle MCD \cong \triangle CMG$

$\therefore \angle MCG = \angle CAB$

We know $\angle PCG = \angle CAB$, so point P is on the diagonal CM of parallelogram $CDMG$. We also know that DG is another diagonal, thus P is the intersection point of the diagonal lines. Therefore, $DP = GP$.

Problem 24

The figure shows right triangle ABC with side lengths 5, 12 and 13. Squares are drawn on each side, and segments DE, FG and HI are drawn between vertices of the squares as shown. What is the area of hexagon $DEFGHI$? (2017 Mathcounts State Target)

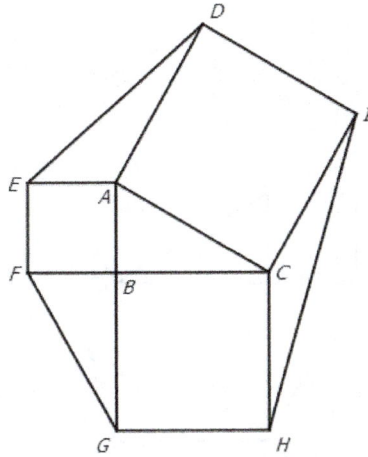

Tips: *Prove that each of the triangles have the same area as triangle ABC.*

Solution 1:

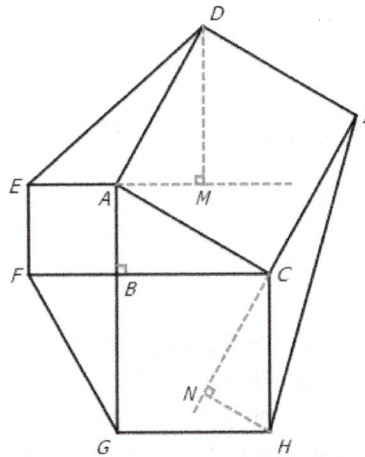

Because (5, 12, 13) is a Pythagorean triplet, $\triangle ABC$ is a right triangle, based on which we can calculate the area of three squares and the $Rt\triangle BCF$.

Let M be on the extension of EA such that $DM \perp EM$. Because $ACID$ is a square, $AC = AD$ and $\angle DAC = 90°$. We also know $\angle BAM = 90°$, thus $\angle BAC = \angle MAD$. Therefore, $\triangle ABC \cong \triangle AMD$ and $DM = BC$.

$\because AE = AB, DM = BC$

$\therefore S_{\triangle}ADE = S_{\triangle}ABC = 30$

Using the same method, we have $\triangle ABC \cong \triangle HNC$ and $S_{\triangle ADE} = S_{\triangle ABC} = 30$. Therefore, the total area of $DEFGHID = 12^2 + 13^2 + 5^2 + 4*30 = 458$.

Solution 2:

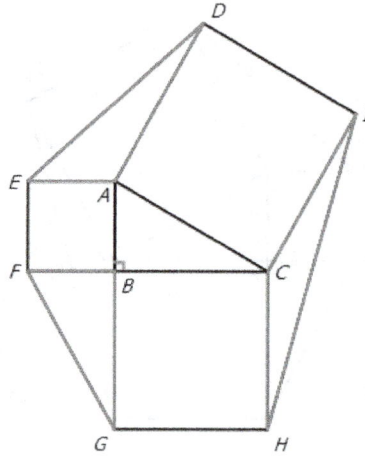

In this solution, we want to prove that $\triangle ADE$, $\triangle FBC$ and $\triangle CHI$ have the same area as $\triangle ABC$.

For $\triangle ADE$, its area is $S_{\triangle ADE} = AB * AD * \sin(\angle DAE)$. With squares we have $AE = AB$, $AD = AC$, and $\angle DAE + \angle BAC = 180°$.

Thus $\sin(\angle DAE) = \sin(\angle BAC)$ and $S_{\triangle ADE} = S_{\triangle ABC}$.

Similarly, we can prove $S_{\triangle FBC} = S_{\triangle ABC}$ and $S_{\triangle CHI} = S_{\triangle ABC}$.

Then the total area of $DEFGHID = 12^2 + 13^2 + 5^2 + 4 * 30 = 458$.

Problem 25

In right triangle ABC, $AB = 5$, $BC = 12$ and $AC = 13$. P lies on side AC so that ray BP bisects angle ABC. Q and R lie on side BC so that $BQ = QR = RC$. Segments AQ and AR intersect segment BP at X and Y, respectively. What is the area of quadrilateral $QRYX$? Express your answer as a common fraction. (2017 Mathcounts State Team)

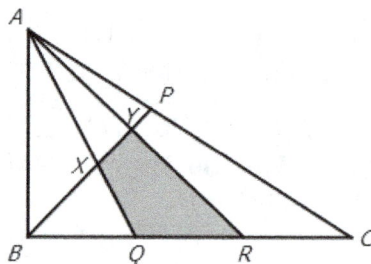

Tips: *1. Use the mass point method or Menelaus's Theorem to get ratios of segments.*
2. Use the area method.

Solution 1:

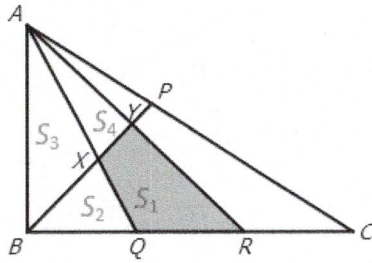

Use the area method with the help of the angular bisector theorem to solve this problem.

The area of four specific shapes is illustrated in the diagram as S_1, S_2, S_3 and S_4. The area of the $Rt\triangle ABC$ is 30. With $BQ = QR = RC$, we have $S_2 + S_3 = 10$ and $S_1 + S_4 = 10$.

Since BP is the angular bisector of angle ABC, we have

$$\frac{S_2}{S_3} = \frac{QX}{XA} = \frac{BQ}{AB} = \frac{4}{5}$$

and

$$\frac{S_1 + S_2}{S_3 + S_4} = \frac{RY}{YA} = \frac{BR}{AB} = \frac{8}{5}$$

Therefore, $S_2 = \frac{40}{9}, S_3 = \frac{50}{9}$, and $S_1 = \frac{920}{117}$.

Thus the area of $QRYX$ is $\frac{920}{117}$.

Solution 2:

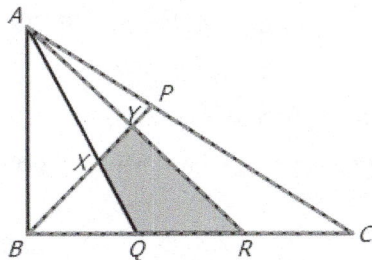

We can use **Menelaus's Theorem** to calculate the ratios of segments to solve this problem.

A is on the extension line of CP and AR crosses two edges of $\triangle BCP$, thus the conditions of **Menelau's Theorem** are met. Using **Menelaus's Theorem**, we have

$$\frac{CA}{PA} * \frac{PY}{YB} * \frac{BR}{RC} = 1$$

Thus

$$\frac{17}{5} * \frac{PY}{YB} * \frac{2}{1} = 1$$

Thus

$$\frac{PY}{YB} = \frac{5}{34}$$

and

$$\frac{PY}{PB} = \frac{5}{39}$$

Using the same method, we can obtain

$$\frac{PX}{PB} = \frac{10}{27}$$

Therefore,

$$S_{\triangle AXY} = \frac{85}{351} S_{\triangle ABP} = \frac{85}{351} * \frac{5}{17} S_{\triangle ABC} = \frac{250}{117}$$

$$S_{QRYX} = S_{\triangle AQR} - S_{\triangle AXY} = \frac{920}{117}$$

Solution 3:

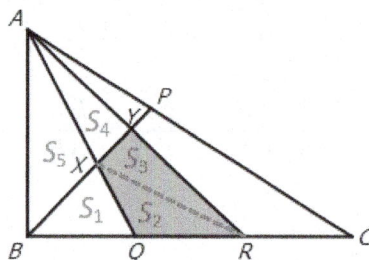

This problem can also be solved using area relationships to calculate the area of $QRYX$.

Since BP is the angular bisector of angle ABC, in $\triangle ABQ$, $S_3/S_2 = 5/4$. Thus $S_3 = \frac{5}{9} * 10 = 50/9$.

In $\triangle ABQ$, $(S_3 + S_4)/(S_1 + S_2) = 5/8$ and $S_3 + S_4 = 5/13 * 20 = 100/13$.

Thus $S_4 = 100/13 - 50/9 = 250/117$ and $S_1 = 10 - 250/117 = 920/117$.

Problem 26

Congruent, non-overlapping circles A, B, C and D are positioned in a plane, such that A, B and C are mutually tangent to each other, and circle D is tangent to circle C. Triangle EFG circumscribes the four circles as shown. If the radius of each circle is 1 meter, then find the length of side FG.(2018 Mathcounts State Sprint)

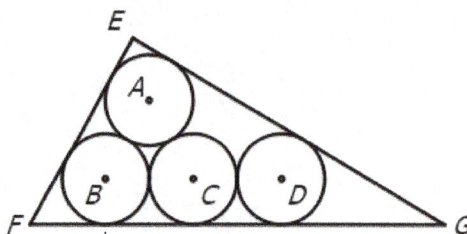

Tips: *1. Generally when segments are tangent to circles, it is useful to connect the centers to the tangent points. 2. Examine what type of triangle $\triangle ABD$ is and its relationship to the triangle EFG.*

Solution 1:

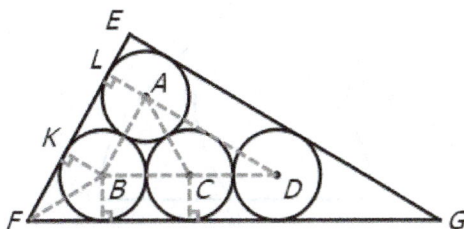

To use the properties of tangent lines, we need to connect tangent points and the centers of circles.

As shown in the diagram, because segment FG is tangent to the three circles of the same diameter, it is evident that $BD // FG$, $AB // EF$ and $AD // EG$.

So $\triangle ABD \sim \triangle EFG$. In $\triangle ABD$, we have $AC = BC = CD = AB = 1$.

Thus $\triangle ABD$ is a right triangle with $\angle ABD = 60°$. The $\triangle EFG$ is also a right triangle with $\angle EFG = 60°$. Then it is easy to show $\triangle KFB$ is a right triangle with $\angle KFB = 30°$. Thus $KF = \sqrt{3}$.

Because $EL = 1$ and $LK = 2$, we know that $EF = 3 + \sqrt{3}$. In the $Rt\triangle EFG$ with $\angle EFG = 60°$, we have $EG = \sqrt{3} * EF = \sqrt{3} * (3 + \sqrt{3}) = 3 + 3\sqrt{3}$.

Therefore, the area of $\triangle EFG$ is $9 + 6\sqrt{3}$.

Problem 27

A figure shown is bounded by segment AB and paths BC and CA, and is formed by seamlessly joining two 45-degree sectors of a circle of radius 1 unit along two sides of a unit square that has C and D as opposite vertices. A quarter-circular arc of radius 1 unit is drawn from C to D. What is the percent probability that a random point chosen on arc CD forms an acute triangle with points A and B? Express your answer to the nearest tenth. (2018 Mathcounts State Sprint)

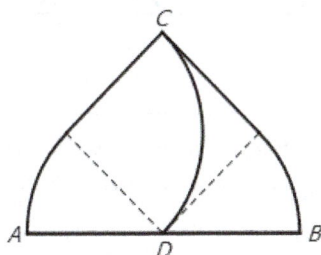

Solution:

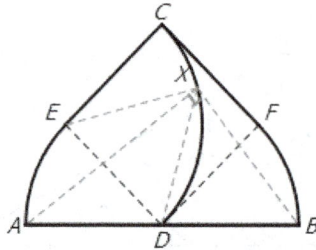

The goal is to find a point X on arc CD to form a right triangle AXB and calculate $\angle XED$.

Draw a point X on arc CD and connect AX, DX, BX and EX. When AXB is a right triangle with D as the midpoint of AB, we have $XD = AD = BD$.

We also know that $EX = ED = AD$, thus $EX = ED = XD$. Since triangle EXD is equilateral, $\angle DEX = 60°$. Any point above X will give an acute triangle and the corresponding arc CX is $30°$.

Therefore, there is a 33.3% chance that a random point on arc AC will form an acute triangle with points A and B.

Problem 28

Diagonal XZ of rectangle $WXYZ$ is divided into three segments each of length 2 units by points M and N as shown. Segments MW and NY are parallel and are both perpendicular to XZ. What is the area of $WXYZ$? Express your answer in simplest radical form. (2016 Mathcounts State Sprint)

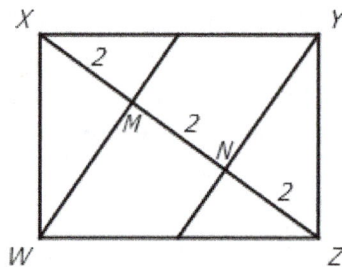

Solution:

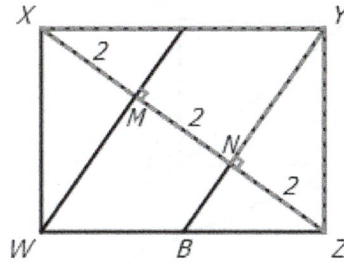

The given information, which involves the altitude to the hypotenuse, suggests the use of a similar triangle pattern in right triangles.

In right $\triangle XYZ$, YN is the altitude of hypotenuse XZ, which leads to $NY^2 = XN * NZ$.

Thus $NY^2 = 4 * 2$ and $NY = 2\sqrt{2}$.

Therefore, the area of $XYZW$ is $XZ * NY = 12\sqrt{2}$.

Problem 29

Rectangle $ABCD$ is shown with $AB = 6$ units and $AD = 5$ units. If AC is extended to point E such that AC is congruent to CE, what is the length of DE? (2016 Mathcounts State Sprint)

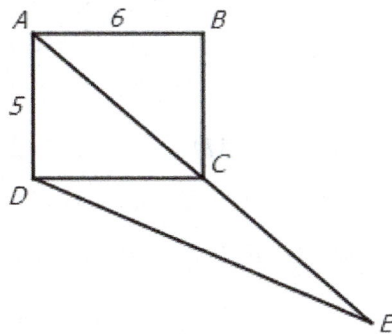

Tips: *Construct a right triangle along the extension line CE, which is similar to the triangles in the rectangle.*

Solution 1:

83

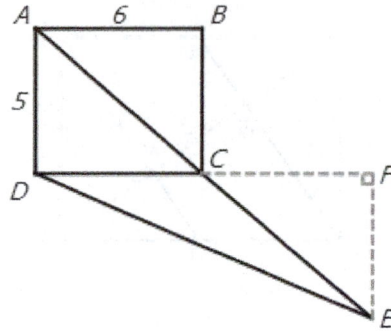

Extend DC to F such that $EF \perp CF$. Because $\angle ADC = \angle EFC = 90°$, $\angle ACD = \angle ECF$ and $AC = CE$, $\triangle ACD \cong \triangle ECF$.

Thus $CF = CD = 6$ and $FE = AD = 5$. In the $Rt\triangle DFE$, using the **Pythagorean Theorem**, we have $DE = 13$.

Solution 2:

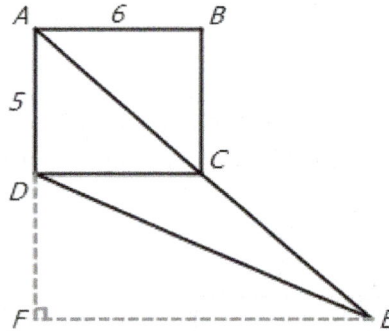

Extend AD to point F such that $EF \perp AD$ to form a $Rt\triangle AFE$.

Because $EF \perp AD$ and $CD \perp AD$, $CD // EF$ and $AD/AF = AC/AE = CD/FE = 2$.

Therefore, in $Rt\triangle DFE$, $DF = 5$ and $EF = 12$, thus $DE = 13$.

Problem 30

A unit square contains four congruent non-overlapping equilateral triangles as shown in the figure. What is the largest possible side-length of one of the triangles? Express your answer as a decimal to the nearest thousandth? (2018 Mathcounts State Team)

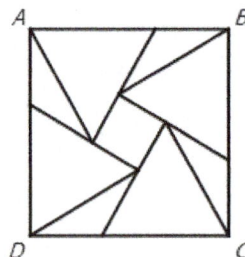

Tips: *Look for the constraint of the diagram for the largest equilateral triangle.*

Solution:

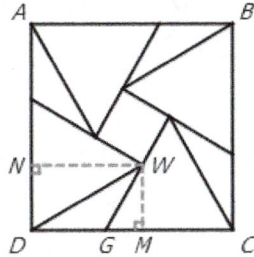

Let points M and N be on CD and AD, respectively, such that $WM \perp CD$ and $WN \perp AD$. In $\triangle DGW$, $\angle GDW = \angle GWD = 30°$. Let $x = GC$. Then, $DW = x$ and $DN = WM = x/2$. In $\triangle GWM$, $\angle GWM = 30°$, thus $WG = \frac{1}{\sqrt{3}}x$.

Because $DG + GC = CD$ and $DG = WG$, we get the equation $x + \frac{1}{\sqrt{3}}x = 1$.

Solving the equation gives us $x = 2\sqrt{3} - 3 \sim 0.634$.

Problem 31

Equilateral triangles ABC and XYZ of side lengths 3 cm and 5 cm, respectively, overlap, as shown, to form equilateral triangle XBP of side length 1 cm. What is the perimeter of concave pentagon $AYZPC$? (2005 Mathcounts State Target)

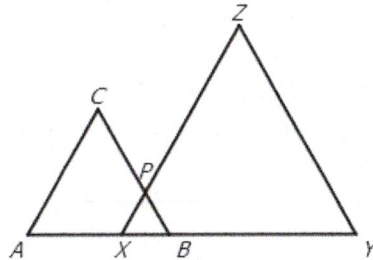

Tips: *Since there are three equilateral triangles here, we just need to carefully add the segments together.*

Solution:

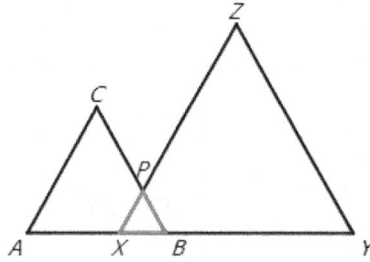

Because $\triangle XPB$ is an equilateral triangle with a side length of 1, $CP = 2, ZP = 3$ and $AY = 7$.

Thus $AY + ZY + ZP + PC + AC = 7 + 5 + 4 + 2 + 3 = 21$ cm.

Problem 32

In equilateral triangle ABC, point D and E are on line BC and BA such that $BD = AE$. Prove $CE = DE$.

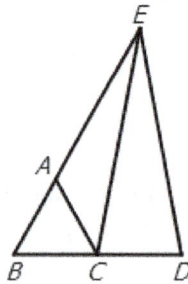

> **Tips:** 1. Put CE and DE in two congruent triangles (construct a triangle through parallel line through point E. 2. Construct congruent triangles using parallel line through point D (parallel to AC). 3. Use the law of cosines to calculate the length of CE and DE and show they are equal.

Solution 1:

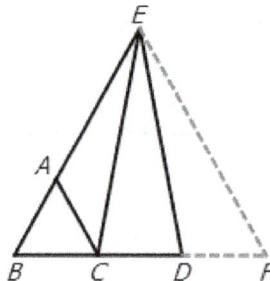

Extend CD to F such that $EF//AC$.

86

$\because \triangle ABC$ is equilateral and $EF // AC$

$\therefore \triangle EBF$ is also equilateral

$\therefore EF = EB = BF, \angle B = \angle F = 60°$

$\because BD = AE$

$\therefore DF = BF - BD = EF - AE = AB = AC$

$\because EF = EB, DF = BC$ and $\angle B = \angle F$

$\therefore \triangle EBC \cong \triangle EFD$

$\therefore CE = DE$

Solution 2:

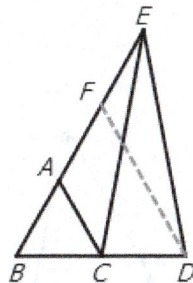

The idea here is similar to solution 1, except that a parallel line is created internally. Let point F be on segment AE such that $DF // AC$.

$\because \triangle ABC$ is equilateral and $DF // AC$

$\therefore \triangle FBD$ is also equilateral

$\therefore FD = BD = BF, \angle B = \angle BFD = 60°$

$\because BD = AE$

$\therefore EF = AE - AF = BF - AF = AB = AC$

$\because EF = AC, AE = FD$ and $\angle CAE = \angle EFD = 120°$

$\therefore \triangle EAC \cong \triangle DFE$

$\therefore CE = DE$

Solution 3:

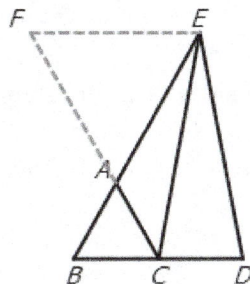

Create another equilateral triangle externally. Extend CA to F such that $EF // BD$.

$\because \triangle ABC$ is equilateral and $EF // BD$

$\therefore \triangle FAE$ is also equilateral

$\therefore FE = FA = AE, \angle B = \angle F = 60°$

$\because AB = AC$

$\therefore FC = AF + AC = AE + AB = EB$

$\because BD = AE$

$\therefore FE = BD$

$\because FC = EB, FE = BD$ and $\angle F = \angle B = 60°$

$\therefore \triangle FEC \cong \triangle BDE$

$\therefore CE = DE$

Solution 4:

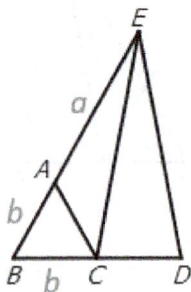

This is a perfect example to use the law of cosines to calculate CE and DE.

Assuming $AE = BD = a$ and $AB = CB = b$, we have $\angle B = 60°$.

Applying the law of cosines on $\triangle EBC$ gives

$$CE^2 = b^2 + (a+b)^2 - 2b(a+b)\cos(60°) = a^2 + b^2 + ab$$

Similarly, applying the law of cosines on $\triangle EBD$, we have

$$DE^2 = a^2 + (a+b)^2 - 2a(a+b)\cos(60°) = a^2 + b^2 + ab$$

Therefore, $CE = DE$.

Problem 33

Point C and D are chosen on the sides of right triangle ABE, as shown, such that the four segments AB, BC, CD and DE each have length 1 inch. What is the measure of angle BAE, in degrees? Express your answer as a decimal to the nearest tenth. (2015 Mathcounts State Sprint)

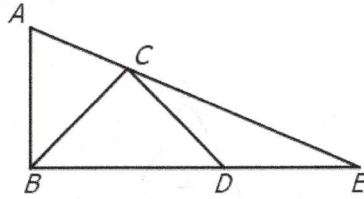

Tips: *Examine the angles in the isosceles triangles and relate them to the right triangle ABE.*

Solution:

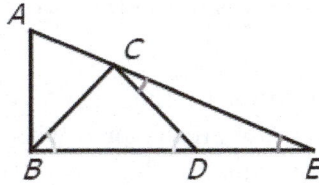

There are three isosceles triangles and one right triangle. Let $\angle E = x$, we have $\angle DCE = x$, $\angle CBD = \angle CDB = 2x$, and $\angle A = \angle ACB = 3x$.

In the $Rt\triangle ABE$, we have $\angle A + \angle E = 90°$. Thus $\angle E = x = 22.5°$.

Problem 34

Circle O is tangent to two sides of equilateral triangle XYZ. If the two shaded regions have areas 50 cm^2 and 100 cm^2 as indicated, what is the ratio of the area of triangle XYZ to the area of circle O? Express your answer as a decimal to the nearest hundredth. (2015 Mathcounts State Target)

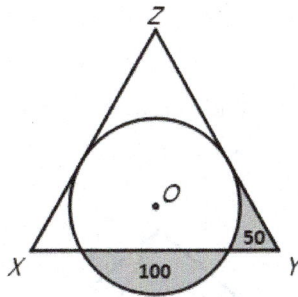

Tips: *1. The figure is symmetrical, thus the area at the left corner is also 50. 2. The area of the circle is equal to the part of the area of the equilateral triangle, which is the key constrain.*

89

Solution:

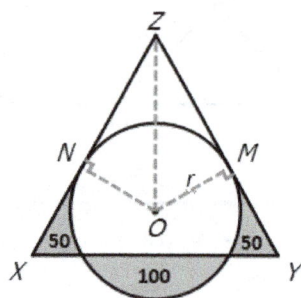

The area of a circle or an equilateral triangle depends on only one parameter (the radius or edge-length). The idea here is to correlate the radius of the circle with the edge length of the equilateral triangle.

Let r be the radius of circle O and a be the edge length of the equilateral triangle XYZ. From the properties of symmetry, the area colored with green is also $50\ cm^2$. Thus the area of $XYMN$ (involving arc MN) is equal to the area of the circle.

The area $XYMN$ with arc MN is

$$S_{\triangle XYZ} - (2S_{\triangle OMZ} - \frac{1}{3}\pi) = \frac{\sqrt{3}}{4}a^2 - (\sqrt{3}r^2 - \frac{1}{3}\pi r^2)$$

The area of circle O is πr^2. Therefore,

$$\frac{\sqrt{3}}{4}a^2 - (\sqrt{3}r^2 - \frac{1}{3}\pi r^2) = \pi r^2$$

Thus the area ratio is

$$\frac{\frac{\sqrt{3}}{4}a^2}{\pi r^2} = \frac{\sqrt{3} + \frac{2}{3}\pi}{\pi} = 1.22$$

Problem 35

In $\triangle ABC$, $\angle ABC = \angle ACB = 40°$ and point P is inside the triangle such that $\angle PBC = 10°$ and $\angle PBC = 20°$. Find angle $\angle PAB$.

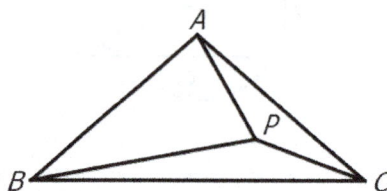

Tips: *1. Use the method of angle chasing by rotating the edges to form triangles.*

Solution 1:

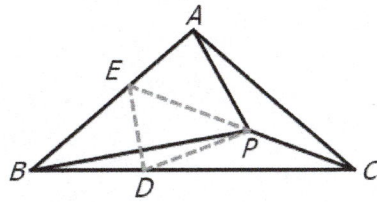

Since this problem involves multiple angles, we need to calculate each angle carefully.

Extend CP and intersect segment AB at point E, and Point D is on segment BC such that $\angle PDC = \angle PCD$, then connect ED.

$\because \angle PDC = \angle PCD = 20°$

$\therefore \angle DPB = \angle DBP = 10°$

$\because \angle EPB = \angle PBC + \angle PCD = 30° = \angle EBP$

$\therefore BDPE$ is a kite shape

$\therefore \angle BED = \angle PED$ and $\angle BDE = \angle PDE = 80°$

$\because \angle AEP = \angle ABC + \angle PCB = 60°$

$\therefore \angle BED = \angle PED = 60° = \angle AEP$

$\therefore \triangle AEC \cong \triangle DEC$

$\therefore AE = DE$

$\because \angle AEP = \angle DEP = 60°$

$\therefore \triangle AEP \cong \triangle DEP$

$\therefore \angle PAB = \angle PDE = 80°$

Solution 2:

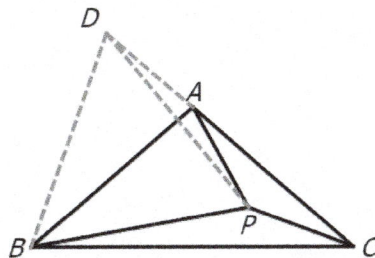

Extend CA to D such that $CD = CB$.

$\because CD = CB$, $\angle DCB = 40°$, PC is the angular bisector of $\angle ACB$.

$\therefore \angle CDB = \angle DBC = 70°$ and CP is the perpendicular bisector of BD.

$\therefore PD = PB$ (isosceles triangle PDB)

$\because \angle PBC = 10°$

$\therefore \angle PBD = 60°$

91

$\therefore \triangle PDB$ is an equilateral triangle

$\because \angle ABP = 30°$

$\therefore \angle ABD = 60° - 30° = 30°$

$\therefore \triangle ABP \cong \triangle ABD$

$\therefore \angle APB = \angle CDB = 70°$

$\because \angle AP = 30°$

$\therefore \angle PAB = 80°$

Solution 3:

Let point E be on the angular bisector of $\angle ABC$ such that $\angle ECD = \angle PCD$. It is straightforward to figure out $\angle EBP = \angle PBC = 10°$ and thus PB is the angular bisector of $\angle EBP$.

$\because PB$ are PC are the angular bisectors of $\angle EBP$ and $\angle DCB$, respectively

$\therefore P$ is the incenter of $\triangle DBC$

$\therefore PD$ is the angular bisector of $\angle CDB$ ($\angle PDB = \angle PDC$)

$\because \angle CDB = 120°$

$\therefore \angle PDC = \angle PDB = 60°$

$\therefore \angle EDC = \angle PDC = 60°$

$\therefore \triangle EDC \cong \triangle PDC$

$\therefore \triangle EDC \cong \triangle PDC \cong \triangle PDC$

$\therefore EC = PC$

$\therefore \triangle AEC \cong \triangle APC$

$\therefore \angle PAC = \angle EAC$

$\because \angle ECA = \angle EBA = 20°$

$\therefore$ Points A, B, C, E are concyclic

$\therefore \angle PAC = \angle EAC = \angle EBC = 20°$

$\therefore \angle PAB = 80°$

Problem 36

In quadrilateral $ABCD$, $\angle A = \angle C = 90°$ and $AD = AB$. If the area of the quadrilateral $ABCD$ is 12, find $BC + CD$.

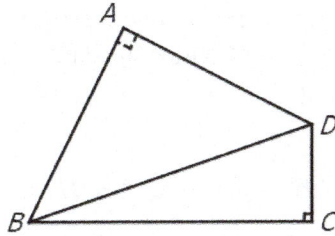

Tips: *Express the area of the quadrilateral as a function of edge-lengths BC and CD.*

Solution:

The area of the isosceles $Rt\triangle ABD$ is $\frac{1}{4}BD^2$ and the area of $Rt\triangle BCD$ is $BC*CD$. Applying the **Pythagorean Theorem** on $Rt\triangle BCD$, we have $BD^2 = BC^2 + CD^2$.

The area of quadrilateral $ABCD$ is equal to

$$\frac{1}{4}(BC^2 + CD^2) + \frac{1}{2}BC*CD = \frac{1}{4}(BC + CD)^2 = 12$$

Thus $BC + CD = 4\sqrt{3}$.

Problem 37

In $\triangle ABC$ $\angle ABC = \angle ACB = 50°$ and point P is inside the triangle such that $\angle PBA = \angle PAB = 10°$, find the angle of $\angle PCB$.

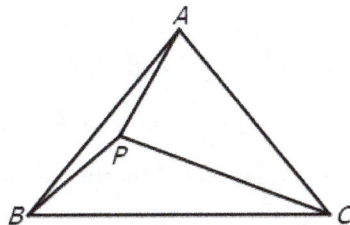

Tips: *1.Construct an equilateral triangle along segment AB or AP. 2. Construct a triangle congruent to triangle APC.*

Solution 1:

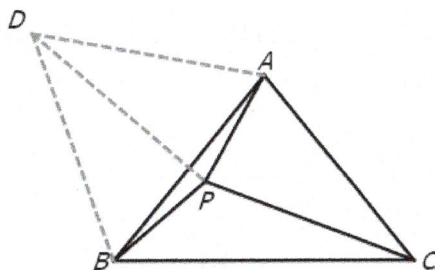

93

To solve these types of problems, it is recommended to first calculate the easy-to-know angles, which will give a clear view of all angle information. It is straightforward to see: $\angle ABC = \angle ACB = 50°$, $\angle PBA = \angle PCA = 10°$, $\angle PAC = 70°$, $\angle BAC = 80°$, $\angle PBC = 40°$, and $AB = AC$.

Let D be a point outside of $\triangle ABC$ such that $\triangle ABD$ is an equilateral triangle.

$\because AD = AC$ and $\angle DAP = \angle CAP = 70°$

$\therefore \triangle DAP \cong \triangle CAP$

$\therefore \angle ACP = \angle ADP$

$\because AD = BD$ and $AP = BP$

$\therefore$ In the kite-shaped $APBD$, DP is the angular bisector of $\angle ADB$

$\therefore \angle ACP = \angle ADP = 30°$

$\therefore \angle PCB = 20°$

Solution 2:

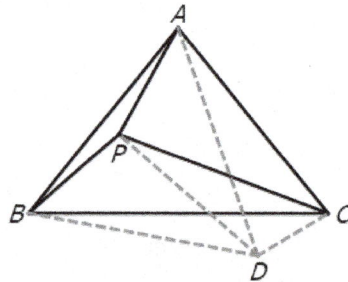

We have obtained these from solution 1: $\angle ABC = \angle ACB = 50°$, $\angle PBA = \angle PCA = 10°$, $\angle PAC = 70°$, $\angle BAC = 80°$, $\angle PBC = 40°$, and $AB = AC$.

Construct an equilateral triangle ABD and connect DP and DC.

We can prove $ACDP$ is concyclic, meaning that the vertices lie on a circle.

$\because \triangle ABC$ is equilateral

$\therefore \angle ADB = \angle DAB = 60°$ and $AD = AB = AC$.

$\therefore \angle CAD = 20°$ and $\angle ADC = 80°$

$\because AP = BP$

$\therefore$ In the kite-shaped $ADBP$, PD is the angular bisector of $\angle ADB$.

$\therefore \angle ADP = 30°$

$\therefore$ In quadrilateral $ACDP$, $\angle ADP + \angle ADC + \angle CAP = 180°$

$\therefore$ Points $ACDP$ are concylic

$\therefore \angle ACP = \angle ADP = 30°$

$\therefore \angle PCB = 20°$

Solution 3:

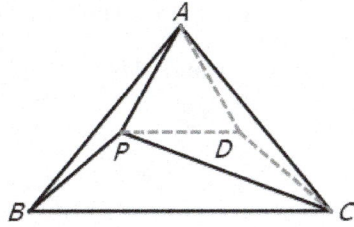

Use the same information obtained in solution 1: $\angle ABC = \angle ACB = 50°$, $\angle PBA = \angle PCA = 10°$, $\angle PAC = 70°$, $\angle BAC = 80°$, $\angle PBC = 40°$ and $AB = AC$.

Construct an equilateral $\triangle APD$ and connect DC.

$\because \triangle APD$ is equilateral

$\therefore \angle ADP = \angle DAP = 60°$ and $AD = AP = PD = DC$

$\therefore \angle CAD = \angle DCA = 10°$ and $\angle ADC = 160°$

$\therefore \angle PDC = 140°$

$\because PD = DC$

$\therefore \angle PCD = 20°$ and $\angle PCA = \angle PCD + \angle DCA = 30°$

$\therefore \angle PCB = 20°$

Problem 38

Triangles ABD and DEF are isosceles right triangles. Points A, D, F and C are colinear, and points B, E and C are colinear. If $AB = BD = 4$ and $ED = EF = 2$, what is the length of segment AC? Express your answer in simplest radical form. (2014 Mathcounts State Sprint)

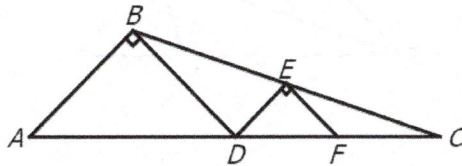

> **Tips:** *1. Find similar triangles to get segment ratios.*

Solution:

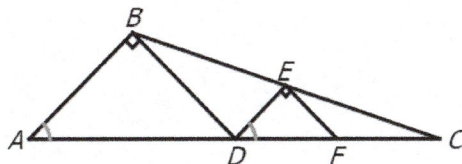

In the two isosceles right triangle ABD and DEF, $\angle BAD = \angle EDF = 45°$.

95

Thus $DE//AB$ and $CD/AC = DE/AB = 1/2$, therefore $CD = AD$.

In isosceles right triangle ABD, $AB = 4$ and $AD = 4\sqrt{2}$.

Therefore, $AC = 2AD = 8\sqrt{2}$.

Problem 39

In $\triangle ABC$, $AB = AC$ and $\angle BAC = 90°$. M is the midpoint of segment AC. D is on segment BC such that $AD \perp BM$. Prove $\angle ADB = \angle MDC$.

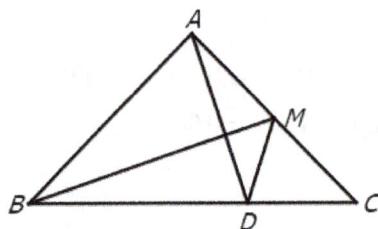

> **Tips:** *1. Prove the angles are equal through similar triangles. 2. Construct a isosceles triangle with both of these angles.*

Solution 1:

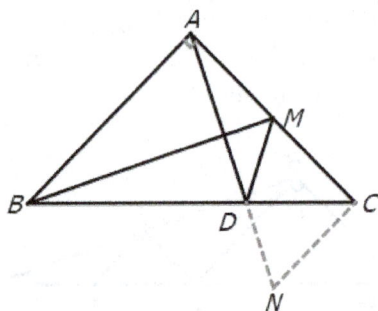

Extend AD to N such that $CN//AB$, and connect NC. We just need to prove $\triangle CMD \cong \triangle CND$.

The first step is to prove $\triangle ABM \cong \triangle CAN$.

In isosceles $Rt\triangle ABC$, $AB = AC$, $\angle BAC = 90°$ and $\angle ABC = \angle ACB = 45°$

$\because CN//AB$

$\therefore \angle ACN = 90° = \angle BAC$ and $\angle DCN = \angle ABC = 45°$

$\because AD \perp BM$

$\therefore \angle CAN = \angle ABM$

$\therefore \triangle ABM \cong \triangle CAN$

96

$$\therefore CN = AM = MC$$

$$\because CN = MC \text{ and } \angle DCN = \angle MCB$$

$$\therefore \triangle MCD \cong \triangle NCD$$

$$\therefore \angle MDC = \angle NDC = \angle ADB$$

Solution 2:

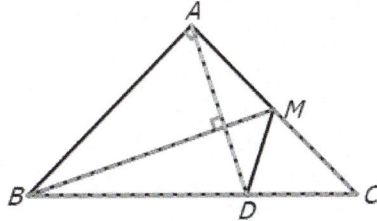

From the conclusion, we can see that if $\triangle MCD \sim \triangle ABD$, we will have $\angle MDC = \angle ADB$. This means we just need to prove $\triangle MCD \sim \triangle ABD$.

$\because$ In the $Rt\triangle ABM$, $AD \perp BM$

$\therefore AB^2 = BE * BM$ and $AM^2 = EM * BM$

$\because AB = AC = 2AM$

$\therefore BE = 4EM$

By the **Menelaus's Theorem,**

$$\frac{AC}{AM} * \frac{EM}{BE} * \frac{BD}{DC} = 1$$

we obtain $BD = 2DC$

$\because \frac{AB}{MC} = \frac{BD}{CD} = 2$ and $\angle ABD = \angle MCD$

$\therefore \triangle ABD \sim \triangle MCD$

$\therefore \angle MDC = \angle ADB$

Solution 3:

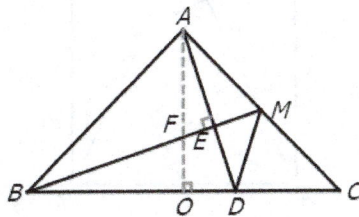

Let point O be on segment BC such that $AO \perp BC$, and AO intersects BM at point F.

$\because \angle AEF = \angle BOF = 90°$, $\angle AFE = \angle BFO$

$\therefore \triangle AFE \sim \triangle BFO$

and

$\because \angle AOD = \angle BOF$ and $AO = BO$

$\therefore \triangle AOD \cong \triangle BOF$

$\therefore OF = OD$

$\therefore AF = CD$

In $\triangle AFM$ and $\triangle CDM$, $AF = CD$, $AM = CM$ and $\angle MAF = \angle MCD = 45°$

$\therefore \triangle AFM \cong \triangle CDM$

$\therefore \angle MDC = \angle AFM = \angle ADB$

Problem 40

In square $ABCD$, each vertex is connected to the midpoints of its two opposite sides, as shown. What is QR/PQ? Express your answer as a common fraction. (2014 Mathcounts State Team)

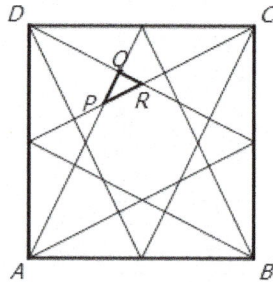

> **Tips:** 1. Create multiple similar triangle pairs to calculate the length of each segment.

Solution:

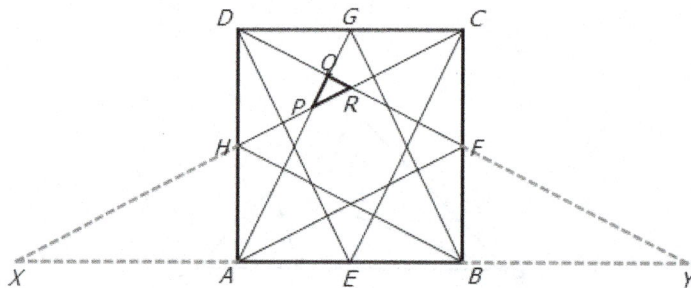

The idea here is to calculate the length of QR and PQ with respect to a standard length (e.g., DF).

Let the length of hypotenuse of the right triangle be a. Extend DF and CH and intersect with segment AB at Y and X, respectively. $QR = DR - QR$ and $PQ = PG - QG$. We know that $DR = 0.5a$.

$\because \triangle DGQ \sim \triangle YAQ$

$\therefore DQ/QY = DG/AY = GQ/AQ = 1/4$

$\therefore DQ = \frac{1}{5}DY = \frac{2}{5}a$ and $GQ = \frac{1}{5}a$

$\therefore QR = \frac{1}{2}a - \frac{2}{5}a = \frac{1}{10}a$

Similarly,

$\because \triangle GCP \sim \triangle AXP$

$\therefore GP/PA = GC/AP = 1/2$

$\therefore GP = \frac{1}{3}GA = \frac{1}{3}a$

$\therefore PQ = \frac{1}{3}a - \frac{1}{5}a = \frac{2}{15}a$

$\therefore QR/PQ = 3/4$

Problem 41

In $\triangle ABC$, $AB = AC$ and D and E are on segment AB and AC such that $BD = CD$. Connect DE and intersect BC at point F. Prove $DF = FE$.

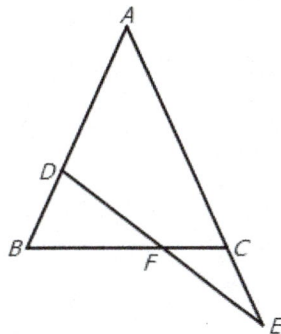

Tips: *1. Construct a congruent triangle pair based on congruent segment information. 2. Construct a parallel line to transfer information to one triangle.*

Solution 1:

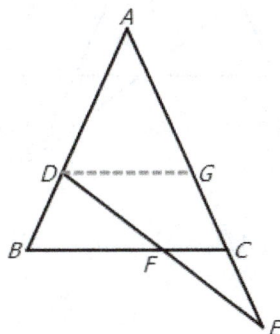

99

Let point G be on segment AC such that $DG // BC$.

$\because AB = AC$ and $DG // BC$

$\therefore GC = BD$ and $DF/FE = GC/CE$

$\therefore DF/FE = BD/CD = 1$

$\therefore DF = FE$

Solution 2:

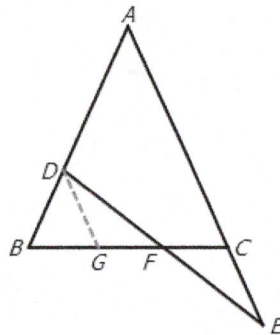

Let point G be on segment BC such that $DG // AC$.

$\because AB = AC$ and $DG // AC$

$\therefore DG = BD$

$\because BD = CE$

$\therefore DG = CE$

$\because DG // AC$

$\therefore \triangle DGF \cong \triangle ECF$

$\therefore DF = FE$

Solution 3:

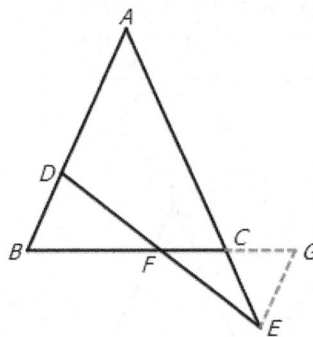

Extend BC to G such that $EG // AB$, and connect CG.

$\because AB = AC$ and $EG // AB$

$\therefore GE = CE$

$\because BD = CE$

$\therefore GE = BD$

Along with $DG // AC$

$\therefore \Delta EGF \cong \Delta DBF$

$\therefore DF = FE$

Solution 4:

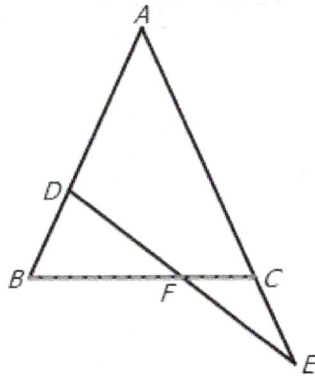

By **Menelaus' theorem**, we have:

$$\frac{AB}{DB} * \frac{DF}{FE} * \frac{CE}{AC} = \frac{AB}{AC} * \frac{CE}{DB} * \frac{DF}{FE} = 1$$

$\because AB = AC$ and $DB = CE$

$\therefore DF = FE$

Problem 42

In ΔABC, $AB = AC$, D is the midpoint of BC, and E is the midpoint of AD. BE and AC intersect at point F. Prove $FC = 2AF$.

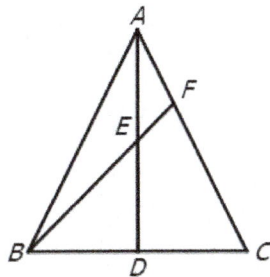

Tips: *1. Construct parallel lines to transfer information into one triangle. 2. Use the mass point method or Menelaus's Theorem.*

Solution 1:

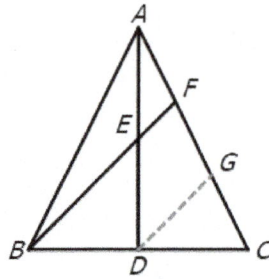

Let G be a point on segment AC such that $DG//BF$.

$\because DG//BF$, $AE = ED$ and $BD = DC$

$\therefore AF = FG$ and $CG = FG$

$\therefore FC = 2AF$

Solution 2:

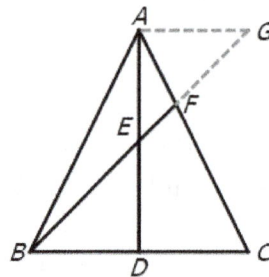

Extend BF to G such that $AG//BC$.

$\because AG//BC$, $AE = ED$

$\therefore AG = BD$

$\because CD = BD$

$\therefore BC = 2AG$

$\because AG//BC$

$\therefore FC/AF = BC/AG = 2$

$\therefore FC = 2AF$

Solution 3:

102

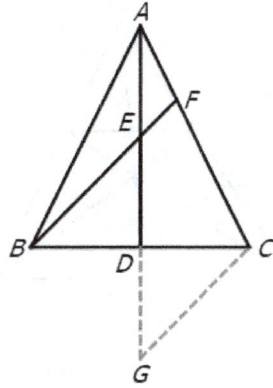

Extend AD to G such that $GC // BF$.

$\because GC // BF$, $BD = CD$

$\therefore DG = ED$

$\because AE = ED$

$\therefore EG = 2AE$

$\because GC // BF$

$\therefore FC/AF = EG/AE = 2$

$\therefore FC = 2AF$

Solution 4:

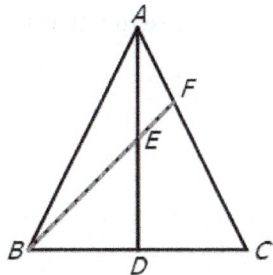

Using **Menelaus's Theorem**, we have:

$$\frac{CB}{DB} * \frac{DE}{EA} * \frac{AF}{FC} = \frac{2}{1} * \frac{1}{1} * \frac{AF}{FC} = 1$$

$\therefore CF = 2AF$

Problem 43

In $\triangle ABC$, D is the midpoint of BC and O is on segment AD. The extensions of BO and CO intersect AC and AB at points E and F, respectively. Prove $FE // BC$.

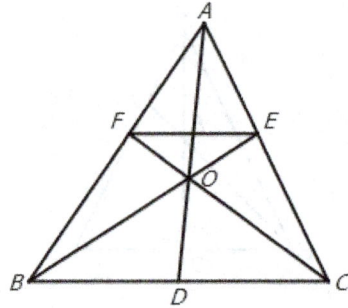

Tips: *1. The way to prove that two lines are parallel is to find an equal ratio of segments. 2. Use Ceva's Theorem to find the ratios of segments.*

Solution 1:

Applying **Ceva's Theorem** in $\triangle ABC$ gives us:

$$\frac{CD}{DB} * \frac{BF}{FA} * \frac{AE}{EC} = 1$$

$\because CD = DB$

$\therefore BF/FA = EC/AE$

$\therefore FE // BC$

Solution 2:

We can also use the area method to find segment ratios.

$\because BD = DC$

$\therefore S_{OAB} = S_{OAC}$

$\because \frac{AF}{FB} = \frac{S_{\triangle OAC}}{S_{\triangle OBC}}$ and $\frac{AE}{EC} = \frac{S_{OAB}}{S_{OBC}}$

$\therefore \frac{AF}{FB} = \frac{AE}{EC}$

$\therefore FE // BC$

Solution 3:

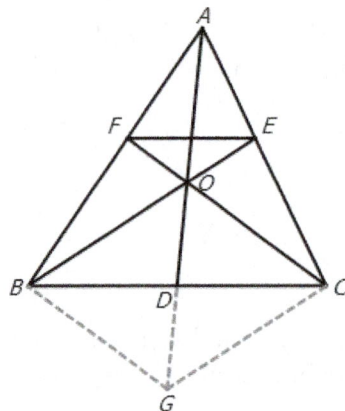

104

Extend AD to G such that $CG//BE$, and connect BG.

$\because BD = DC$ and $CG//BE$

$\therefore OCGB$ is a parallelogram

$\therefore OC//BG$

$\therefore \frac{AF}{FB} = \frac{AO}{OG} = \frac{AE}{EC}$

$\therefore FE//BC$

Problem 44

In $\triangle ABC$, $\angle ACB = 60°$ and $\angle BAC = 75°$. $AD \perp BC$ at D and $BE \perp AC$ at E. AD and BE intersect at H. Find $\angle CHD$.

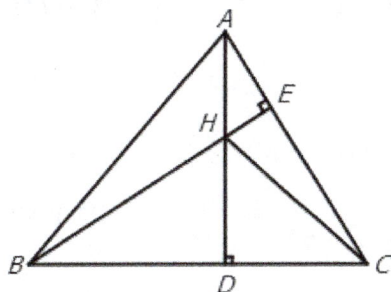

> **Tips:** *Prove triangle BDH is congruent to triangle ADC.*

Solution 1:

The idea here is to prove $\triangle HDC$ is an isosceles right triangle.

$\because \angle ACB = 60°$, $\angle BAC = 75°$

$\therefore \angle ABD = 45°$

$\because AD \perp BC$

$\therefore \angle BAD = 45°$ and $BD = AD$

$\because AD \perp BC$ and $BE \perp AC$

$\therefore \triangle BDH \cong \triangle AEH$

$\therefore \angle DBH = \angle DAC$

$\therefore \triangle BDH \cong \triangle ADC$

$\therefore HD = CD$

$\therefore \angle CHD = 45°$

Solution 2:

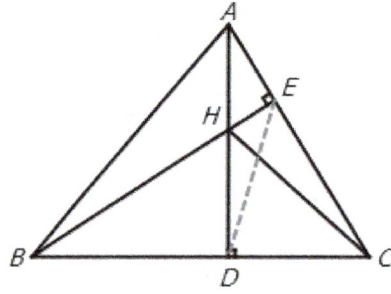

Connect points D and E.

$\because AD \perp BC$ and $BE \perp AC$

$\therefore ABDE$ and $CDEH$ are concyclic

$\therefore \angle CHD = \angle CED = \angle ABD$

It is straightforward to find $\angle ABD = 45°$. Thus $\angle CHD = 45°$.

Problem 45

In $\triangle ABC$, $AC = BC = 5$, $ACB = 80°$. O is a point inside $\triangle ABC$. If $OAB = 10°$, $OBA = 30°$, find the length AO.

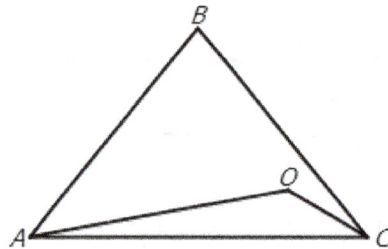

> **Tips:** 1. Reflect point O across segment AC to point F and connect AF, OF and CF.

Solution 1:

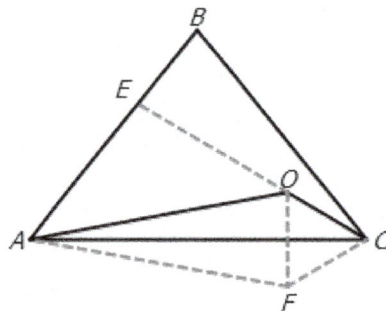

To solve these types of problems, it is helpful to first list all of the angle information provided.

We have $\angle B = 80°$, $\angle BAO = 40°$, $\angle OAC = 10°$, $\angle BCA = 20°$, and $\angle OCA = 30°$.

Extend CO to intersect AB at point E. Rotate CO to CF such that $\angle OCF = 60°$ and connect OF and AF.

Because $\angle EOA = 40° = \angle BAO$, and $\angle BEC = 80° = \angle B$, $OE = AE$, and $CE = BC = AB$

$\therefore EB = OC$

It is straightforward to see that $\triangle OCF$ is an equilateral triangle and AC is the angular bisector of $\angle OCF$.

$\therefore OF = OC = EB$ and $\triangle OAC \cong \triangle EAC$

$\therefore AF = AO$ and $\angle OAF = 20°$

$\because$ In $\triangle OAF$ and $\triangle BCE$, $AF = BE$, $CE = BC$, $AF = AO$, and $\angle OAF = \angle BCE = 20°$

$\therefore \triangle OAF \cong \triangle BCE$

$\therefore OA = BC = 5$

Solution 2:

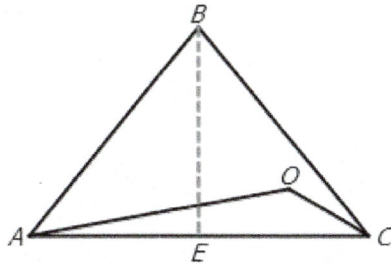

Another useful method is trigonometry. Let E be the midpoint of segment AC, and connect BE. Since $AB = CB$, BE is the perpendicular bisector of AC. It is straightforward to find the following angle information: $\angle ABE = 40°$, $\angle OCA = 30°$, and $\angle AOC = 140°$.

In $\triangle OAC$, by the law of sines,

$$\frac{OA}{AC} = \frac{\sin \angle OCA}{\sin \angle AOC} = \frac{\sin 30°}{\sin 140°} = \frac{\sin 30°}{\sin 40°}$$

In $\triangle ABC$, we have $AC = 2AE = 2AB * \sin 40°$.

Thus $OA = AC * \sin 30° / \sin 40° = 2AB * \sin 30° = AB = 5$.

Problem 46

In $\triangle ABC$, $AD \perp BC$ at D and $BE \perp AC$ at E. AD and BE meet at F. If $BF = AC$, find $\angle ABC$.

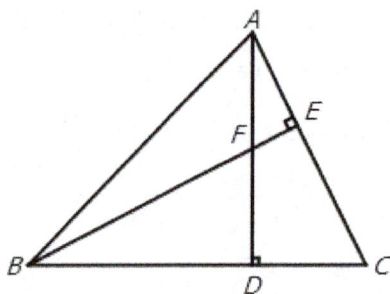

Tips: *Prove that triangle BFD is congruent to triangle ACD, and triangle ABD is an isosceles right triangle.*

Solution:

In $\triangle ABC$, $AD \perp BC$ and $BE \perp AC$

$\therefore \angle AEF = \angle ADB = 90°$

$\therefore \triangle AEF \sim \triangle BDF$

$\therefore \angle EAF = \angle DBF$

$\therefore \triangle ADC \sim \triangle BDF$

$\because BF = AC$

$\therefore \triangle ADC \cong \triangle BDF$

$\therefore AD = BD$

$\therefore \angle ABC = 45°$

Problem 47

In the isosceles right $\triangle ABC$, $\angle A = 90°$ and points E and F on hypotenuse BC such that $\angle EAF = 45°$. Prove: the triangle formed by three sides BD, DE and EC is a right triangle.

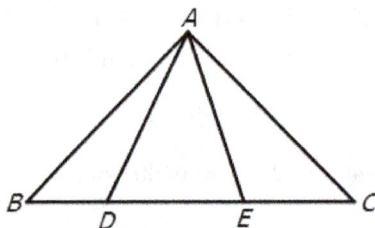

Tips: *Transfer segments BD, DE and EC into one triangle and prove that it is a right triangle.*

Solution 1:

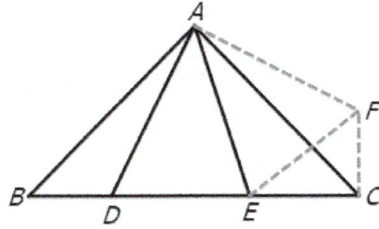

To prove that the triangle formed by the three segments is a right triangle, we need to put these three segments in one triangle and then prove the triangle formed is a right triangle using the **Pythagorean theorem**.

Since $AB = AC$ and $\angle DBA = 90°$, we can rotate $\triangle ABD$ around point A by $90°$ to create $\triangle ACF$, and connect EF.

Thus $AF = AD$, $FC = BD$, $\angle FAE = 45°$, and $\angle FCA = \angle B = 45°$.

$\because AF = AD$, $\angle FAE = \angle DAE = 45°$

$\therefore \triangle FAE \cong \triangle DAE$

$\therefore FE = DE$

In $\triangle FCE$, $FC = BD$, $FE = DE$ and $\angle FCE = 90°$. Thus the three segments BD, DE and EC form a right triangle.

Solution 2:

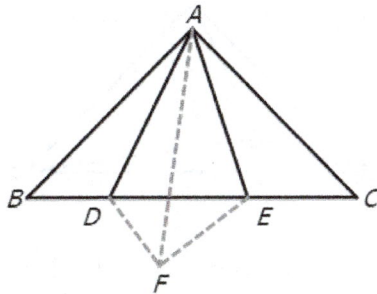

Reflect $\triangle ACE$ along the segment AE to the other side to create $\triangle AEF$, and connect DF.

$\because \triangle AEF \cong \triangle AEC$ (reflecting creates congruent triangles)

$\therefore AF = AC$, $EF = CE$ and $\angle AFE = \angle C = 45°$, and $\angle CAE = \angle FAE$

$\because \angle EAD = 45°$ and $\angle BAC = 90°$

$\therefore \angle BAD + \angle CAE = 45° = \angle EAD$

$\because \angle CAE = \angle FAE$

$\therefore \angle DAF = 45° - \angle FAE = 45° - \angle CAE = \angle BAD$

$\because AB = AC = AC, \angle DAF = \angle BAD$

$\therefore \triangle ABD \cong \triangle AFD$

$\therefore DF = BD$ and $\angle AFD = \angle B = 45°$

In $\triangle FDE$, because $DF = BD$, $FE = EC$ and $\angle DFE = 90°$, the three segments BD, DE and EC form a right triangle.

Problem 48

As shown in the figure, $\triangle ABC$ is an isosceles right triangle. Point P is inside $\triangle ABC$ and $PC = 13$, $PB = 5$, and $PA = 2\sqrt{6}$. Find the length of AB.

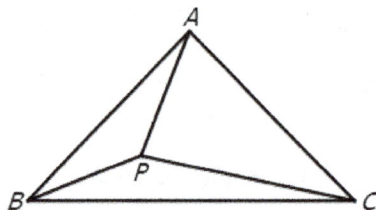

Tips: *Rotate triangle ABP or ACP to form a new triangle containing all the side length information.*

Solution 1:

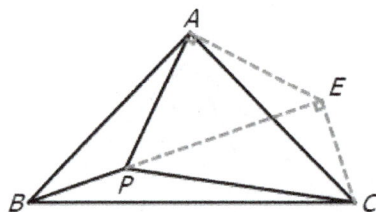

Rotate $\triangle ABP$ around point A by $90°$ to create ACE. Thus $\triangle APE$ is also an isosceles triangle and $AE = AP$ and $EC = BP = 5$, thus $PE = 12$.

In $\triangle PEC$, $PE = 12$, $EC = 5$ and $PC = 13$. Using converse of the **Pythagorean theorem**, we know $\triangle PEC$ is a right triangle and $\angle PEC = 90°$.

In $\triangle AEC$, $AE = 6\sqrt{2}$, $EC = 5$ and $\angle AEC = 135°$. Using the law of cosines, we have

$$AC^2 = 72 + 25 + 2 * 6\sqrt{2} * 5 * \cos 135° = 157$$

Thus $AB = AC = \sqrt{157}$.

Solution 2:

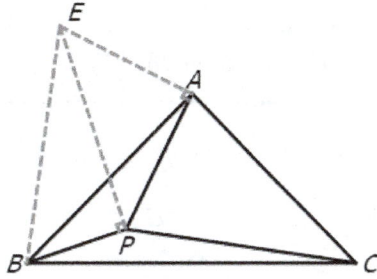

Using a similar approach as solution 1, we rotate $\triangle ABP$ $90°$ clockwise around point A. In $\triangle APB$, we have $\angle APB = 135°$ and AB can be calculated through the law of cosines.

Solution 3:

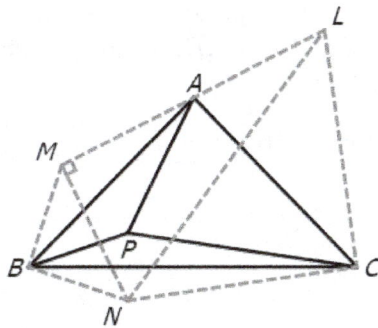

Reflect $\triangle APB$, $\triangle BPC$ and $\triangle CPA$ along AB, BC and AC, respectively, to create three pairs of congruent triangles. It is straightforward to see that $\angle MAL = 180°$ (M, A and L are conlinear), $\angle MBN = 90°$ and $\angle NCL = 90°$.

So $LN = 13\sqrt{2}$, $MN = 5\sqrt{2}$ and $ML = 12\sqrt{2}$.

By the converse **Pythagorean Theorem**, $\triangle MNL$ is a right triangle and $\angle NML = 90°$.

The total area of $BNCLM$ is the sum of the area of three right triangles, which is 314. On the other hand, the area of $BNCLM$ is twice as much as the area of $\triangle ABC$, which is $\frac{1}{2}AB^2$. Therefore $AB = \sqrt{157}$.

Problem 49

In the $\triangle ABC$, $AB = AC$ and $\angle A = 100°$. BE is the angle bisector of $\angle ABC$. Prove $AE + BE = BC$.

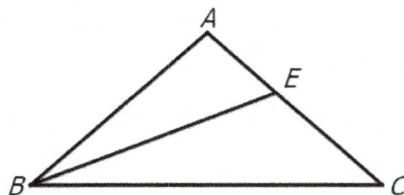

111

Tips: *Cut BC into two segments such that one of the segments is congruent to BE, and prove the other segment is congruent to AE.*

Solution 1:

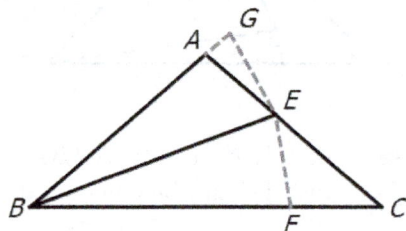

Let point F be on segment BC such that $BF = BE$ and point G be on the extension of BA such that $BG = BE$. Both $\triangle BEG$ and $\triangle BFE$ are isosceles triangles and $\triangle BEG \cong \triangle BFE$. Thus $\angle FEC = 40° = \angle C$ and $\angle G = 80° = \angle EAG$.

$\therefore \triangle CEF$ and $\triangle EAG$ are isosceles triangles

$\therefore GE = AE$ and $EF = FC$

$\because \triangle BEG \cong \triangle BFE$

$\therefore GE = EF$

$\therefore AE = FC$

$\because BE = BF$, $AE = FC$

$\therefore AE + BE = BC$

Solution 2:

Trigonometry can also be used to efficiently solve this problem. In $\triangle AEB$, using the law of sines, we have

$$\frac{AE}{BE} = \frac{\sin \angle ABE}{\sin \angle BAE} = \frac{\sin 20°}{\sin 100°} = \frac{\sin 20°}{\sin 80°}$$

$$\frac{AE + BE}{BE} = \frac{\sin 20°}{\sin 80°} + 1 = \frac{\sin 20° + \sin 80°}{\sin 80°} = \frac{2 \sin 50° \cos 30°}{2 \sin 40° \cos 40°} = \frac{\cos 30°}{\sin 40°}$$

In $\triangle EBC$, using the law of sines, we have

$$\frac{BC}{BE} = \frac{\sin \angle BEC}{\sin \angle ECB} = \frac{\sin 120°}{\sin 40°} = \frac{\sin 60°}{\sin 40°} = \frac{\cos 30°}{\sin 40°}$$

$\therefore \frac{AE + BE}{BE} = \frac{BC}{BE}$

$\therefore AE + BE = BC$

Problem 50

In $\triangle ABC$, $AB = AC$ and $\angle A = 20$. D is on segment AB such that $AE = BC$. Find $\angle BDC$.

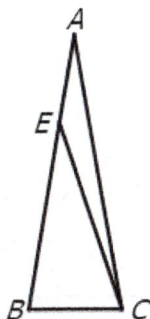

Tips: *Construct an equilateral triangle with one side AC and use information from congruent triangles.*

Solution:

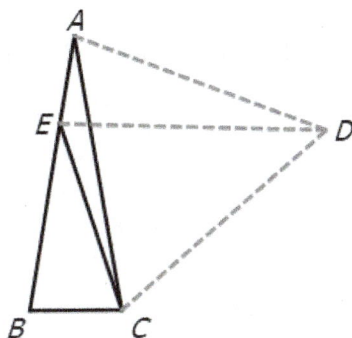

Construct an equilateral triangle ACD along segment AC and connect DE.

$\because AB = AC$ and $\angle A = 20°$

$\therefore \angle B = 80°$

$\because AD = AC = AB$, $AE = BC$ and $\angle DAE = \angle B = 80°$

$\therefore \triangle DAE \cong \triangle ABC$

$\therefore DE = AC$ and $\angle DEA = 80°$

$\because DE = AC = AD$, and $\angle EDC = 60° - 20° = 40°$

$\therefore \angle DEC = 70°$

$\therefore \angle BEC = 180° - 80° - 70° = 30°$

Problem 51

In equilateral $\triangle ABC$, D and E are points on AB and AC such that $AD = CE$. BE and CD intersect at point F. Find $\angle BFC$.

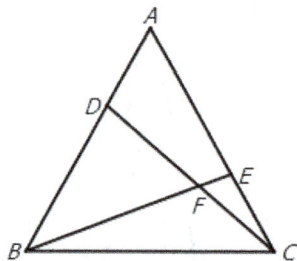

Tips: *1. Prove triangles ADC and CEB are congruent.*

Solution:

In the equilateral $\triangle ABC$, $AC = BC$ and $\angle A = \angle ACB = 60°$.

Together with $AD = CE$, we have $\triangle ADC \sim \triangle CEB$. Thus $\angle ACD = \angle EBC$.

Therefore, $\angle FBC + \angle FCB = \angle ACD + \angle FCB = 60°$. Thus $\angle BFC = 120°$.

(We can also use $\triangle BDC \cong \triangle BEA$ to calculate the angle.)

Problem 52

In $\triangle ABC$, M is the midpoint of side BC and N is on the angular bisector of $\angle BAC$ such that $CN \perp AN$. If $AC = 12$ and $AB = 18$, find MN.

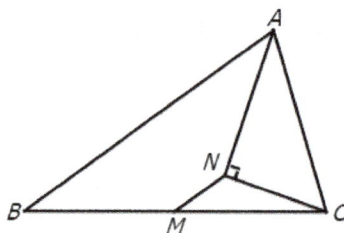

Tips: *1. The angular bisector and perpendicular information suggests the construction of an isosceles triangle. 2. Find a triangle where MN is the midline.*

Solution:

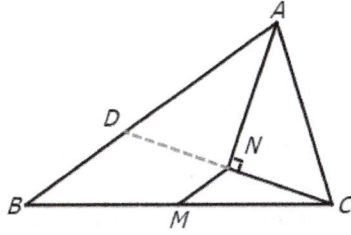

Extend segment CN to meet AB at point D. Because $CN \perp AN$ and AN is the angular bisector of $\angle BCA$, $\triangle ADC$ is an isosceles triangle.

Thus $CN = DN$ and $BD = AB - AD = AB - AC = 18 - 12 = 6$.

In $\triangle CBD$, $CN = DN$ and $CM = BM$.

Therefore, MN is a midline and $MN = BD/2 = 6/2 = 3$.

Problem 53

In $Rt\triangle ABC$, $C = 90°$ and D is on segment BC such that AD is the angle bisector of $\angle BAC$. If $CD = 15$, $AC = 30$, find the length of AB.

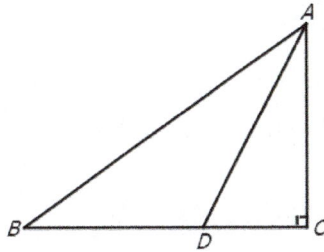

Tips: *Use angular bisector information to construct a pair of congruent triangles.*

Solution 1:

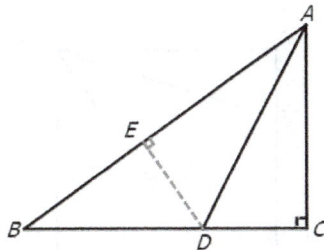

Let point E be on segment AB such that $DE \perp AB$. Let $x = BE$ and $y = BD$. Because AD is the angular bisector of $\angle BAC$ and $DE \perp AB$, $DC \perp AC$, $\triangle AED \cong \triangle ACD$.

115

So $AE = AC = 30$ and $DE = DC = 15$. Applying the **Pythagorean Theorem** gives $(x + 30)^2 = (y + 15)^2 + 30^2$.

It is easy to know $\triangle BDE \sim \triangle BAC$, thus

$$\frac{BE}{ED} = \frac{BC}{AC} \Rightarrow \frac{x}{15} = \frac{y + 15}{30}$$

Solving the equation set gives $x = 20$ and $y = 25$.

Therefore, $AB = 20 + 30 = 50$.

Solution 2:

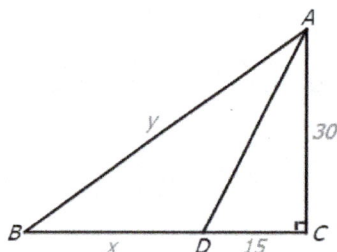

We can also use trigonometry to solve this problem easily. We have $\angle BAC = 2\angle DAC$.

Based on the definition of tangents, we know $\tan \angle DAC = \frac{15}{30} = \frac{1}{2}$

Thus

$$\tan \angle BAC = \tan(2\angle DAC) = \frac{2\tan \angle DAC}{1 - (\tan \angle DAC)^2} = \frac{4}{3}$$

Thus $BC = AC * \tan \angle BAC = 40$ and $AB = 50$.

Problem 54

In square $ABCD$, M and N are on BD and BC such that AN is the angular bisector of $\angle MAC$. Prove: $DM + CN = AM$.

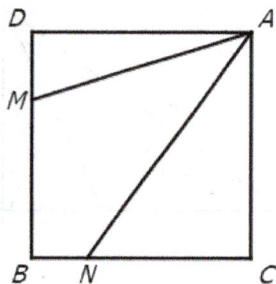

116

Tips: *Rotate triangle ADM counter-clockwise 90 degree around point A to put the target segments into a single triangle.*

Solution 1:

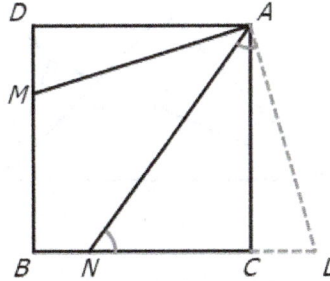

Rotate $\triangle ADM$ 90° counter-clockwise around point A to create $\triangle ACL$. We have $CL = DM$, $AL = AM$ and $\angle CAL = \angle DAM$.

Thus $\angle LAN = \angle CAL + \angle CAN = \angle DAM + \angle MAN = 90° - \angle CAN = \angle ANL$, and $\triangle ANL$ is an isosceles triangle and $AL = NL = AM$. Therefore, $DM + CN = AM$.

Solution 2:

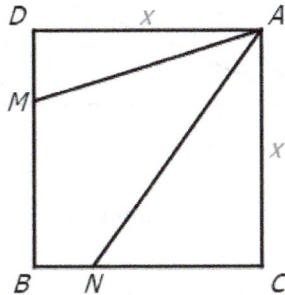

In this solution we use the method of trigonometry to solve the problem. Let the edge-length of the square be x and $\angle DAM$ be a. We know that $\angle CAN = 45° - a/2$.

In $\triangle DAM$, we have $DM = x \tan a$ and $AM = x/\cos a$.

In $\triangle CAN$, we have $NC = x \tan(45° - a/2)$.

Therefore,

$$DM + NC = x \tan a + x \tan\left(45° - \frac{a}{2}\right) = x\frac{\sin\left(45° + a/2\right)}{\cos a \cos\left(45° - a/2\right)} = \frac{x}{\cos a} = AM$$

Problem 55

In $\triangle ABC$, $AC = BC$ and $\angle A = 90°$. D is a point on AC and E is on the extension of BD such that $CE \perp BE$. If $BD = 2CE$. Prove: BD is the angle bisector of $\angle ABC$.

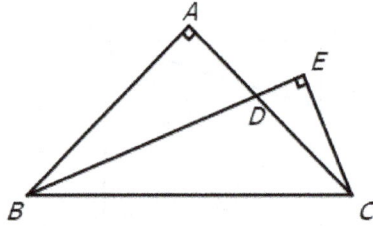

> **Tips:** *1. Extend segments BA and CE to create a complete triangle.*

Solution 1:

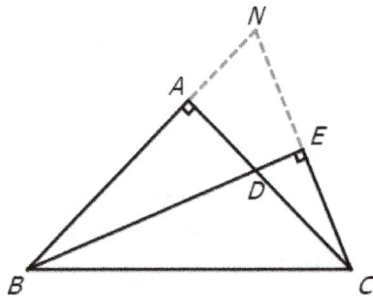

Extend segments CE and BA to meet at point N.

$\because CE \perp BE$ and $\angle BAC = 90°$

$\therefore \triangle ABD \sim \triangle DCE$

$\therefore \angle ABD = \angle DCE$

$\because AB = AC$, $\angle BAD = \angle CAN = 90°$ and $\angle ABD = \angle DCE$

$\therefore \triangle ABD \sim \triangle ACN$

$\therefore BD = CN$

$\because BD = 2CE$

$\therefore NE = CE$

$\because NE = CE$ and $CE \perp BE$

$\therefore BE$ is the perpendicular bisector of segment NC

$\therefore BE$ is the angular bisector of $\angle ABC$

Solution 2:

118

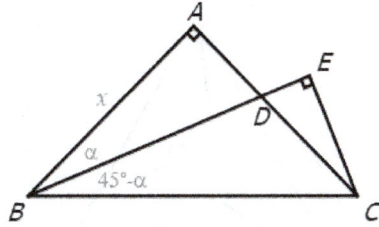

To use trigonometry to solve this problem, first assume $AB = AC = x$, $\angle ABE = \alpha$, $\angle OCA = 45° - \alpha$, and $BC = x\sqrt{2}$.

In $\triangle ABD$, we have $BD = \frac{x}{cos(\alpha)}$. In $\triangle BEC$, we have $CE = BC * \sin(45 - \alpha) = x\sqrt{2} * \sin(45 - \alpha)$.

$\because BD = 2CE$

$\therefore x = 2x\sqrt{2}\sin(45° - \alpha)\cos\alpha$

$\therefore 2\sqrt{2}\sin(45° - \alpha)\cos\alpha = 1$

Expand $\sin(45° - \alpha)$ as $(\sin 45 * \cos\alpha - \cos 45 * \sin\alpha)$ (Difference formula)

$\therefore 2(\cos\alpha)^2 - 2\sin\alpha\cos\alpha = 1$

$\therefore \cos(2\alpha) = \sin(2\alpha)$ (Double angle formula)

Thus $2\alpha + 2\alpha = 90°$ and $\alpha = 22.5° = 45° - \alpha$.

Therefore, BE is the angular bisector of $\angle ABC$.

Problem 56

$\triangle ABC$ is an equilateral triangle and O is a point inside $\triangle ABC$ such that $\angle AOB = 125°$. $\angle AOC = 115°$. Find all the angles of a triangle formed by the three sides OA, OB, and OC.

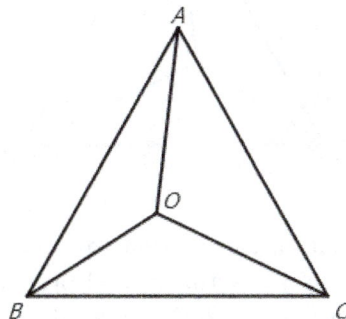

> **Tips:** *The information given inside the equilateral triangle suggests the construction of external equilateral triangles along edges.*

Solution 1:

119

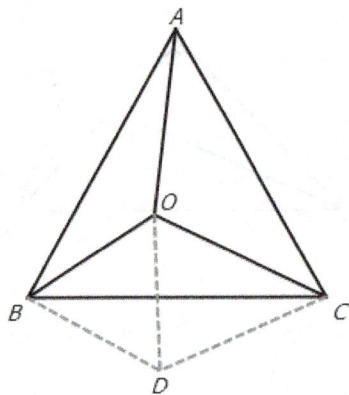

Rotate segment OB 60° clockwise around point B to form an equilateral triangle OBD, and connect OD and DC.

$\because \angle AOB = 125°$, $\angle AOC = 115°$ and $\angle BOD = 60°$,

$\therefore \angle BOC = 120°$ and $\angle DOC = 60°$

$\because \angle ABC = \angle BOD = 60°$

$\therefore \angle ABO = \angle CBD$

$\therefore \triangle CBD \cong \triangle ABO$

$\therefore DC = OA$ and $\angle ODC = \angle BDC - \angle ODB = \angle AOB - 60° = 65°$

$\therefore \angle OCD = 55°$

Therefore, in the triangle formed by OA, OB and OC, the angles are 60°, 65° and 65°, respectively.

Solution 2:

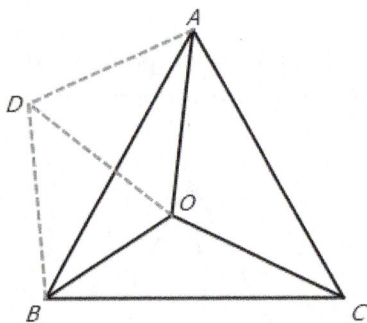

Similarly, a rotation transformation can be done around point A or C, and the process will be the same as in solution 1. Although the special properties of the 120° angle are not used here, it will be helpful to pay attention to this special angle in general.

Problem 57

In $\triangle ABC$, $\angle BAC = 60°$, $AB = 2AC$. Point O is inside $\triangle ABC$ such that $OA = \sqrt{3}$, $OB = 5$, and $OC = 2$. Find the area of $\triangle ABC$.

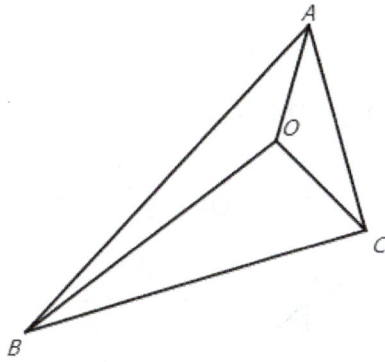

Tips: *Reflect point O across sides AB, BC, and AC and connect all the points.*

Solution:

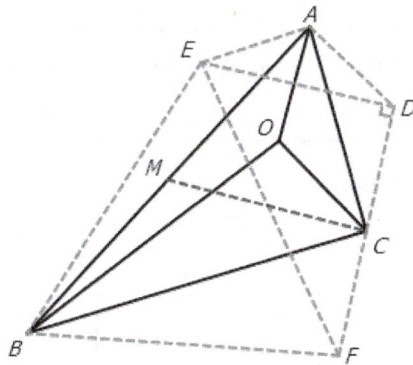

For a very challenging problem like this, the first step is to prove $\triangle ABC$ is a right triangle. Let M be the midpoint of side AB.

$\because \angle BAC = 60°$ and $AM = BM = AC$

$\therefore \angle ACM = 60°$ and $\angle MCB = 30°$

$\therefore \angle ACB = 90°$ and $\angle MBC = 30°$

The second step is to reflect all three triangles inside to the outside as shown in the diagram.

In $\triangle AED$, $AE = AD = OA = \sqrt{3}$ and $\angle EAD = 120°$

In $\triangle EDF$, $CD = CF = OC = 2$ and $\angle FCD = 180°$

In $\triangle BEF$, $BE = BF = OB = 5$ and $\angle FCD = 60°$

Thus in $\triangle AED$, $ED = AD * \sqrt{3} * ED = 3$, and the area is $3\sqrt{3}/4$. $\triangle EDF$ is an equilateral triangle with edge-length of 5, thus its area is $25\sqrt{3}/4$.

In $\triangle EDF$, $ED = 3$, $DF = 4$ and $EF = 5$, thus its area is 6.

The area of polygon $ADCFBE$ is $6 + 7\sqrt{3}$. Therefore, the area of $\triangle AED$ is $3 + 7\sqrt{3}/2$.

Problem 58

The side-length of a square $ABCD$ is 4. M is on BC such that $CM = CD/3$. What is the length of OC?

(A) $4\sqrt{2}$ (B) $2\sqrt{2}$ (C) $\sqrt{2}$ (D) 2   (E) 1/4

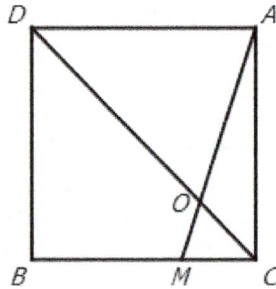

> **Tips:** *Prove triangle OAD is similar to triangle OMC and use the ratio properties of similar triangles.*

Solution:

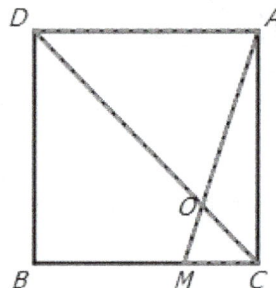

In square $ABCD$, since $AD//BC$ and $AD = BC = 4$, $DC = 4\sqrt{2}$ and $\triangle OAD \sim \triangle OMC$.

Therefore, $OC/OD = CM/AD = 1/3$ and $OC = DC/4 = \sqrt{2}$.

Answer is (C).

Problem 59

In the $Rt\triangle ABC$, AD is the altitude on the side BC. If $AC = 30$ and $BD = 32$, what is the length of CD?

(A) 12   (B) 14   (C) 16   (D) 18   (E) 24

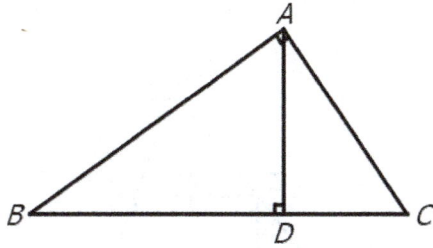

Tips: *Use the property AC*AC = CD*BC*

Solution 1:

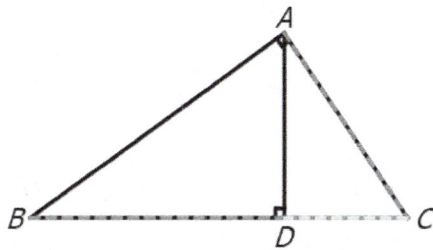

In a right triangle involving the altitude of its hypotenuse, we have the formula $AC^2 = CD * BC$.

Let $x = CD$, we have $30^2 = x * (x + 32)$. Solving this quadratic equation gives $CD = x = 18$.

The answer is (D).

Problem 60

Square $DEFG$ is inscribed in $\triangle ABC$. The areas of $\triangle AOR$, $\triangle BOP$, $\triangle CRQ$ are 1, 3, and 1, respectively. What is the area of square $DEFG$?

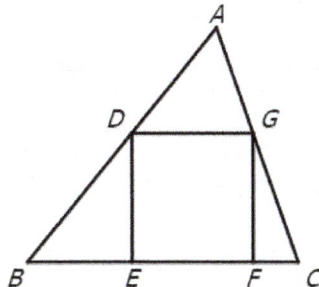

Tips: *Construct the altitude to BC and use the ratio properties of similar triangles (ABC ~ ADG)*

Solution 1:

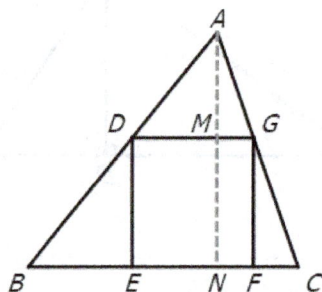

A square brings the implicit condition parallel lines, which can be used to find similar triangles. Let AN be the altitude to segment BC and point M be the intersection of DG and AN.

$\because DG // BC$

$\therefore \triangle ADG \sim \triangle EBC$

$\therefore AM/AN = DG/BC$

Let a be the edge-length of the square, thus $AM = 2/a$, $BE = 6/a$ and $FC = 2/a$, and

$$\frac{2/a}{2/a + a} = \frac{a}{6/a + 2/a + a}$$

Thus $16/(a^2) + 2 = 2 + a^2$ and $a^2 = 4$, which is the area of the square.

Solution 2:

Similar to solution 1, first construct the altitude AN on segment BC and let point M be the intersection of DG and AN. Then we focus on the relationship between the areas of a similar triangle pair.

$\because DG // BC$

$\therefore \triangle ADG \sim \triangle EBC$

$\therefore S_{\triangle ADG}/S_{\triangle ADG} = (DG/BC)^2$

Let the edge-length of the square be a and we have $BE = 6/a$ and $FC = 2/a$.

Thus the area ratio is

$$\frac{1}{(5 + a^2)} = [\frac{a}{(a + 6/a + 2/a)}]^2$$

Simplifying this equation, we have $a^6 + 4a^4 - 16a^2 - 64 = 0$. Factoring the equation with alternate terms, we get $(a^4 - 16)(a^2 - 4) = 0$, so $a^2 = 4$. Therefore, the area of square $DEFG$ is 4.

Problem 61

Rectangle $DEFG$ is inscribed in $\triangle ABC$. If the area of $\triangle ABC$ is 2, what is the largest area of the rectangle $DEFG$?

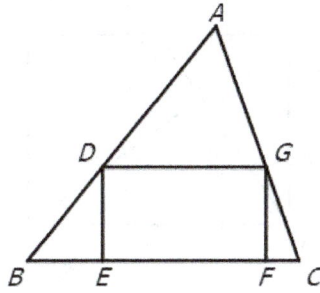

Tips: *Express the area of the rectangle as a function of its width and use algebra to find the maximum area.*

Solution:

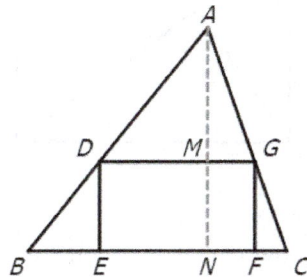

Let AN be the altitude on side BC and intersect with DG at point M. Let $x = DG$. Since $DEFG$ is an inscribed rectangle, $DG // BC$ and $\triangle ADG \cong \triangle ABC$.

$\therefore DG/BC = AM/AN$

$\therefore AM = DG/BC \, AN$

$\therefore MN = AN - AM = AN - DG/BC \, AN$

The area of $DEFG$ is $DGMN = x * AN * (1 - x/BC) = AN * BC * (x/BC) * (1 - x/BC)$, which has the largest value of $AN * BC/2$ when $x/BC = 1 - x/BC$ (or $x = BC/2$).

Therefore, the largest area of $DEFG$ is 1 (half of the area of $\triangle ABC$) when DG is a midline.

Problem 62

In rectangle $ABCD$, $AB = 50\sqrt{6}$ and E is the midpoint of AD. If lines AC and BE are perpendicular, what is the length of AD?

125

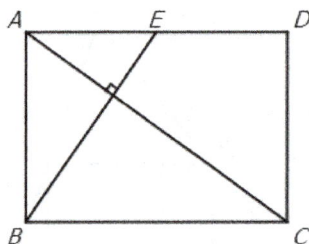

Tips: *Find pairs of similar right triangles and form an algebraic equation for AD.*

Solution:

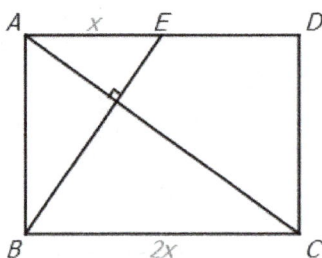

The key constraint is $AC \perp BE$, which leads to multiple pairs of similar triangles. Let $AE = x$ and $BC = 2x$.

Because $\triangle ABE \sim \triangle BAC$, $AE/AB = AB/BC$

Thus $2x^2 = AB^2$, which leads to $AD = 2x = \sqrt{2}AB = 100\sqrt{3}$.

Problem 63

In rectangle $ABCD$, we have $AB = 8$, $BC = 9$, H is on BC with $BH = 6$, E is on AD with $DE = 4$, line EC intersects line AH at G, and F is on line AD with $GF \perp AF$. What is the length GF? (2003 AMC10A)

(A) 16   (B) 20   (C) 24   (D) 28   (E) 30

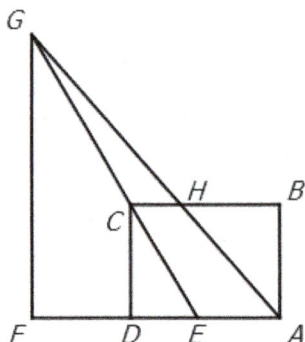

Tips: *Understand that G is determined by two intersecting lines and construct two pairs of similar triangles involving the two lines.*

Solution 1:

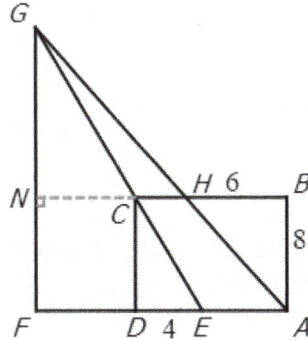

The position of point G is determined by two lines which originate from a rectangle (indicating good chance to form similar triangle pairs).

Extend BC to meet GF at point N. Let NC and GN be x and y. Since $ABCD$ is a rectangle, we have two pairs of similar triangles: $\triangle ABH \cong \triangle GNH$ and $\triangle CDE \cong \triangle GNC$.

We have $GN/AB = NH/BH$ and $GN/CD = NC/DE$, which lead to $y/8 = (x+3)/6$ and $y/8 = x/4$.

Solving the equation set gives $x = 6$ and $y = 12$.

Therefore, $GF = GN + NF = 12 + 8 = 20$.

The answer is (B).

Solution 2:

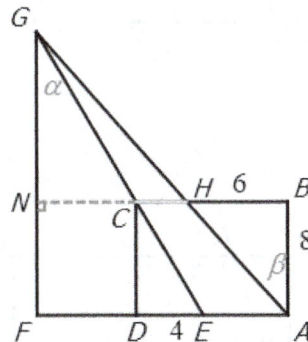

We will now use trigonometry to solve this problem. Assume $\angle CGN = \alpha$, $\angle BAH = \beta$ and $GN = x$.

$\because AB // CD // GF$

$\therefore \angle HGN = \angle BAH = \beta$, $\angle CGN = \angle ECD = \alpha$

127

$\therefore HN = GN\tan\beta = x * \frac{3}{4}$ and $CN = GN\tan\alpha = x * \frac{4}{8}$

$\because CH = 3 = HN - CN$

$\therefore x * \frac{3}{4} - x * \frac{4}{8} = 3$

$\therefore GN = x = 12$

$\therefore GF = GN + NF = 12 + 8 = 20$

The answer is (B).

Problem 64

In rectangle $ABCD$, $AB = 5$ and $BC = 3$. Points F and G are on CD so that $DF = 1$ and $GC = 2$. Lines AF and BG intersect at point E. What is the area of $\triangle AEB$? (2003 AMC10B)

(A) 10   (B) 21/2   (C) 12   (D) 25/2   (E) 15

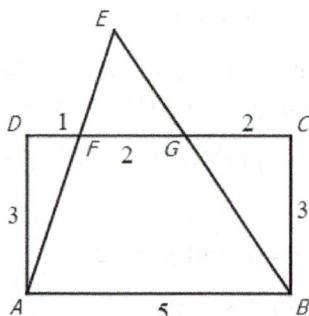

Tips: *1. Construct the altitude of triangle EFG on FG. 2. Find two pairs of similar triangles and use ratio properties to get equations for lengths.*

Solution 1:

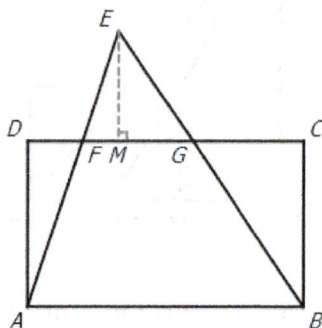

Let M be a point on FG such that $EM \perp FG$. We know that $FG = 2$.

$\because EM // AD // CB$

128

$\therefore \triangle EFM \sim \triangle ADF$ and $\triangle EGM \sim \triangle BGC$

$\therefore EM/AD = FM/DF$ and $EM/CB = MG/GC$

$\therefore FM/DF = GM/CG$

$\because DF = 1, CG = 2$ and $FM + MG = 2$

$\therefore FM = (2 - FM)/2$

$\therefore FM = 2/3$

$\therefore EM = FM/DFAD = 2/3/13 = 2$

Therefore, the area of $\triangle AEB$ is $5 * (2 + 3)/2 = 25/2$.

The answer is (D).

Solution 2:

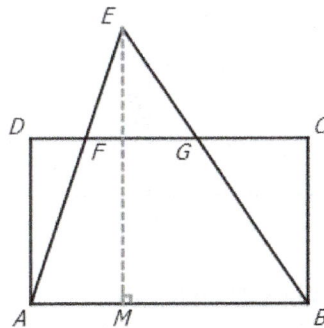

Similarly, let M be a point on FG such that $EM \perp FG$. We know that $FG = 2$.

$\because EM // AD // CB$

$\therefore \triangle EAM \sim \triangle ADF$ and $\triangle EBM \sim \triangle BGC$

$\therefore EM/AD = FM/DF$ and $EM/CB = MG/GC$

$\therefore AM/DF = BM/CG$

$\because DF = 1, CG = 2$ and $AM + BM = 5$

$\therefore AM = (5 - AM)/2$

$\therefore AM = 5/3$

$\therefore EM = AM/DFAD = 5/3/13 = 5$

Therefore, the area of $\triangle AEB$ is $55/2 = 25/2$.

The answer is (D).

Problem 65

$ABCD$ is a square and A, E, F and G are in the same line. Find FG if $AE = 5$ cm and $EF = 3$ cm.

(A) 16/3   (B) 25/3   (C)   5/3   (D) 5   (E) 11/2

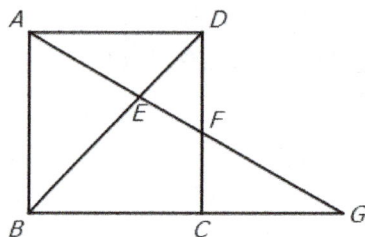

Tips: *1. Discover relations from pairs of similar triangles. 2. Be comfortable to introduce new variables (i.e. side length of the square here).*

Solution 1:

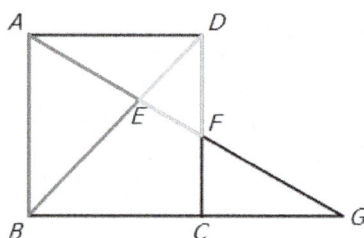

The key is to use pairs of similar triangles involving AE, EF and FG. Different choices of similar triangle pairs, however, affects the complexity of the solution. Here we use the following pairs of similar triangles for simplicity: $\Delta EDF \sim \Delta EBA$ and $\Delta GFC \sim \Delta GAB$.

$\Delta EDF \sim \Delta EBA$ leads to $DF/AB = EF/AE = 3/5$, and $CF/AB = 2/5$.

$\Delta GFC \sim \Delta GAB$ leads to $FG/AG = CF/AB = 2/5$.

Thus $FG/(FG + 8) = 2/5$ and $FG = 16/3$.

The answer is (A).

Solution 2:

We may choose pairs of similar triangles other than those in solution 1, such as $\Delta EAD \sim \Delta EGB$ and $\Delta FAD \sim \Delta FGC$. However, the algebra will be slightly more complex. Let $x = FG$. $\Delta EAD \sim \Delta EGB$ leads to $BG/AD = EG/AE = (3 + x)/5$.

Similarly, $\Delta FAD \sim \Delta FGC$ leads to $CG/AD = FG/AF = x/8$.

With $BG = BC + CG$, we have $BG/AD = 1 + CG/AD = 1 + x/8 = (3 + x)/5$.

Solving the equation above gives $x = FG = 16/3$.

The answer is (A).

Problem 66

Let $ABCD$ be a parallelogram. Extend DA through A to a point P, and let PC meet AB at Q and DB at R. Given that $PQ = 735$ and $QR = 112$, find RC. (1998 AIME)

130

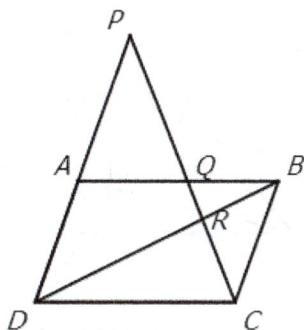

Tips: *Pick similar triangle pairs along PQRC and construct an equation set of ratios.*

Solution 1:

In this solution we use the following pairs of similar triangles: $\triangle RBQ \sim \triangle RDC$ and $\triangle PAQ \sim \triangle PDC$.

$\triangle RBQ \sim \triangle RDC$ leads to $QB/DC = QR/RC$. Similarly, $\triangle PAQ \sim \triangle PDC$ leads to $AQ/DC = PQ/PC = PQ/(PQ + QR + RC)$.

With $QB + AQ = DC$, we have $QR/RC + PQ/(PQ + QR + RC) = 1$. Thus $112/RC + 735/(735 + 112 + RC) = 1$.

Solving the equation above gives $RC = 308$.

Solution 2:

The following two pairs of similar triangles are used: $\triangle RPD \sim \triangle RCB$ and $\triangle QPA \sim \triangle QCB$. Let $RC = x$.

$\triangle RPD \sim \triangle RCB$ leads to $PD/BC = (PQ + QR)/RC = (735 + 112)/x$. $\triangle RPD \sim \triangle RCB$ leads to $PA/BC = PQ/QC = 735/(112 + x)$.

With $PD = PA + AD = PA + BC$, we have $PD/BC = 1 + 735/(112 + x) = (735 + 112)/x$.

Simplifying the equation results in $x^2 = 112 \times 735$. Therefore, $x = RC = 308$.

Problem 67

In $\triangle ABC$, points D and E lie on BC and AC, respectively. Suppose that AD and BE intersect at T so that $AT/DT = 3$ and $BT/ET = 4$. What is the value of CD/BD? (2004 AMC10B)

(A) 1/8   (B) 2/9   (C) 3/10   (D) 4/11   (E) 5/12

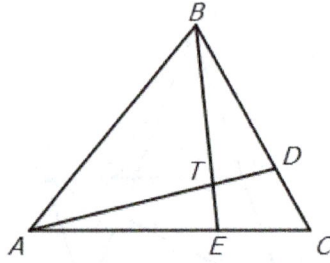

Tips: *This type of problem can be solved by the area method, the mass point method, or constructing parallel lines to shift ratio information.*

Solution 1:

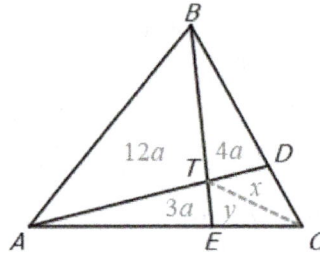

Here we use the area method to solve the problem. Connect TC and let the area of $\triangle TAE$, $\triangle TEC$ and $\triangle TDC$ be $3a$, y and x, respectively. The area of $\triangle TAB$ and $\triangle TBD$ are $12a$ and $4a$, respectively. Using the area method we have the following ratios:

$$\frac{12a}{4a + x} = \frac{3a}{y}$$

and

$$\frac{12a}{3a + y} = \frac{4a}{x}$$

Solving the equation set above, we have $x = 16/(11a)$. Thus $CD/BD = [16/(11a)]/(4a) = 4/11$.

The answer is (D).

Solution 2:

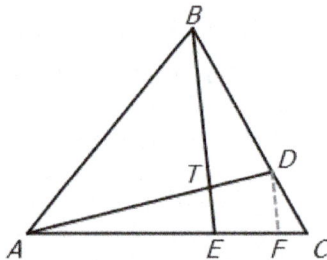

132

Let F be a point on AC such that $DF//BE$.

$\because DF//BE$

$\therefore DF/TE = AD/AT = 4/3$

$\because BT = 4TE$

$\therefore DF/BE = DF/TE \cdot TE/BE = 4/3 \cdot 1/5 = 4/15$

$\because DF//BE$

$\therefore CD/BC = DF/BE = 4/15$

$\therefore CD/BD = 4/11$

The answer is (D).

Solution 3:

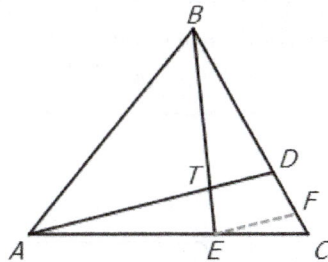

Let F be a point on BC such that $EF//AD$.

$\because EF//AD$

$\therefore EF/TD = EB/TB = 5/44$

$\because AT = 3TD$

$\therefore EF/AD = EF/TD \cdot TD/AD = (5/4) * (1/4) = 5/16$

$\because EF//AD$

$\therefore CD/CF = AD/EF = 16/5$ and $CD/DF = 16/11$

$\therefore CD/BD = CD/DF \cdot DF/BD = (16/11) * (1/4) = 4/11$

The answer is (D).

Solution 4:

Despite solutions 1-3, the mass point method is the most straightforward way to solve this problem.

Put a mass of 1 at point D. Then, the mass at A and T are $1/3$ and $4/3$, respectively. On line BE, the mass at B is $(1/5) * (4/3) * (4/15)$. on line BC, the mass on point C is $11/5$. Thus $CD/BD = 4/11$.

The answer is (D).

Problem 68

In parallelogram $ABCD$, point M is on AB so that $AM/AB = 17/1000$, and point N is on AD so that $AN/AD = 17/2009$. Let P be the point of intersection of AC and MN. Find AC/AP. (2009 AIME)

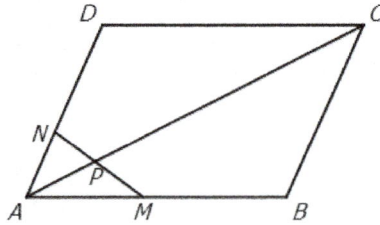

Tips: *Three methods can be used for solving these types of problems: the area method, the mass point method, and constructing parallel lines.*

Solution 1:

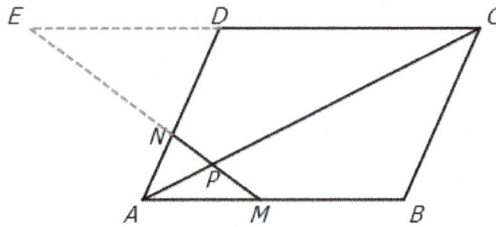

This solution involves constructing parallel lines to find similar triangles. Extend MN and CD to intersect at point E. $ABCD$ is a parallelogram. Thus $AB//CD$ and $AB = CD$. Therefore, we have $\triangle NED \sim \triangle NMA$ and $\triangle PEC \sim \triangle PMA$.

$\triangle NED \sim \triangle NMA$ leads to $AM/ED = AN/ND = 17/1992$; $\triangle PEC \sim \triangle PMA$ leads to $AM/EC = AP/PC = AM/(ED+CD) = 1/(ED/AM+CD/AM) = 1/(1992/17+1000/17)$.

Thus $AC/AP = 1 + PC/AP = 1 + 1992/17 + 1000/17 = 177$.

Solution 2:

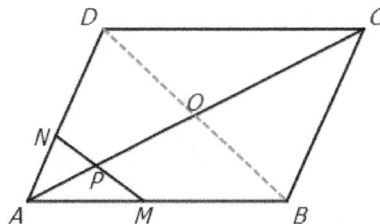

Connect diagonal DB to intersect AC at point O. In the triangle ABD, we can use the mass point method to calculate the ratio of AP/PO first.

O is the midpoint of BD. The mass on O, D, B are 2, 1 and 1, respectively. Along segment AD, $AN/AD = 17/2009$ leads to $AN/ND = 17/1992$.

The mass contribution from DA for point A is 1992/17. Along segment BA, $AM/AB = 17/1000$ leads to $AM/MB = 983/17$, thus the mass contribution from BA for point A is 983/17.

The total mass at point A is 175. Therefore, $AP/PO = 2/175$, so $AC/AP = 2AO/AP = 2(OP/AP + 1) = 177$.

Solution 3:

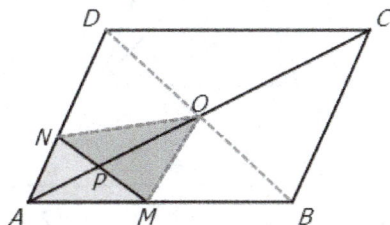

Use the area method to calculate AP/PO first. Connect DB and label the intersection with AC point O. Connect NO and MO. In a parallelogram the area of $\triangle OAD$ and $\triangle OAB$ are equal. Let $S_{\triangle OAD} = S_{\triangle OAB} = s$.

Because $AN/AD = 17/2009$ and $AM/AB = 17/1000$, $S_{\triangle OAN} = (17/2009) * s$, $S_{\triangle OAM} = (17/1000) * s$, and $S_{\triangle ANM} = (17/2009) * (17/1000) * (2s)$. Thus $S_{\triangle ONM} = S_{\triangle OAN} + S_{\triangle OAM} - S_{\triangle ANM} = [(17/2009) + (17/1000) - (17/2009) * (17/1000) * 2] * s$.

Therefore, $AP/PO = S_{\triangle ANM}/S_{\triangle ONM} = 2/177$ and $AC/AP = 2 * (1 + PO/AP) = 177$.

Solution 4:

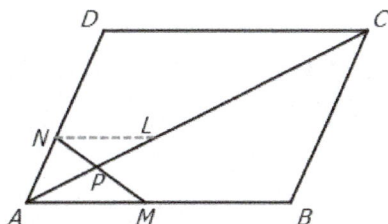

Let L be a point on diagonal AC such that $NL//AB$. $NL//AB$ gives $AL/AC = NL/CD = AN/AD = 17/2009$.

Thus $AP/PL = AM/NL = (AM/AB) * (AB/NL) = (17/1000) * (2009/17) = 2009/1000$ and $AP/AL = 2009/3009$

Therefore, $AP/AC = (AP/AL) * (AL/AC) = (2009/3009) * (17/2009) = 17/3009$, and $AC/AP = 177$.

Problem 69

In $\triangle ABC$, $AB = BC$ and D is a point on AB such that $AC = CD$. If $AB = 12$ and $AC = 5$, what is AD?

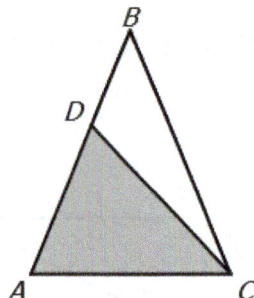

Tips: *1. Find a pair of similar triangles and use their ratio properties.*

Solution:

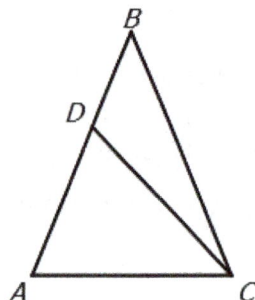

Since $AB = BC$ and $AC = CD$, $\triangle ABC \sim \triangle ACD$, which leads to $AB/AC = AC/AD$. Therefore, $AD = AC^2/AB = 25/12$.

Problem 70

Rectangle $ABCD$ is inscribed in triangle EFG such that side AD of the rectangle is on side EG of the triangle, as shown. The triangle's altitude to side EG is 7 inches, and $EG = 10$ inches. The length of segment AB is equal to half the length of segment AD. What is the area of rectangle $ABCD$? Express your answer as a common fraction. (2007 Mathcounts National Team)

136

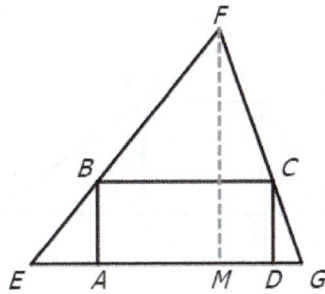

Tips: *Use the ratio information from similar triangles FBC and FEG.*

Solution:

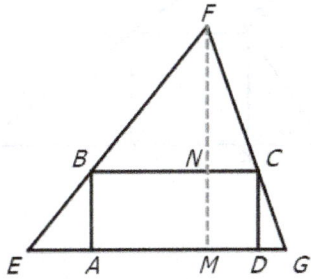

Let $x = AB$. Then, $BC = 2AD = 2x$. In rectangle $ABCD$, $BC // AD$, thus $\triangle FBC \sim \triangle FEG$.

Therefore, $FN/FM = BC/EG \Rightarrow (FM - x)/FM = 2x/EG$, thus $(7 - x)/7 = 2x/10$. Solving the equation above gives us $x = 35/8$, which means the area of $ABCD$ is $38\frac{9}{32}$.

Problem 71

Square $BCFE$ is inscribed in right triangle AGD, as shown below. If $AB = 28$ and $CD = 58$, what is the area of square $BCFE$?

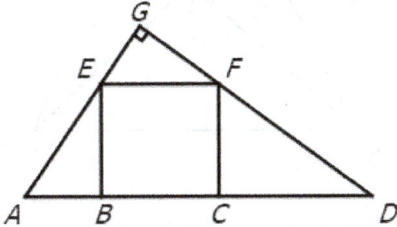

Tips: *Lookout for pairs of similar triangles such as ABE, EGF, and FCD.*

Solution 1:

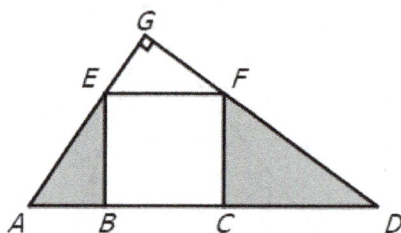

We can easily see that $\triangle ABE \sim \triangle CFD$, thus $BE/CD = AB/FC$. Because $BE = FC$, $AB = 28$ and $CD = 58$, $BE * FC = AB * CD = 1624$, which is the area of square $BCFE$.

Solution 2:

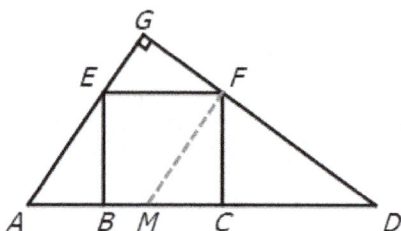

Let point M be on AD such that $FM//AG$. We know that $\triangle FCM \cong \triangle EBA$, thus $MC = AB = 28$. In the $Rt\triangle FMD$, FC is the altitude on the hypotenuse. Therefore, $FC^2 = MC * CD = 1624$, which is the area of square $BCFE$.

Problem 72

Convex quadrilateral $ABCD$ has $AB = 9$ and $CD = 12$. Diagonals AC and BD intersect at E, $AC = 14$, and $\triangle AED$ and $\triangle BEC$ have equal areas. What is AE? (2009 AMC12A)

(A) 6   (B) 7   (C) 5   (D) 4   (E) 12

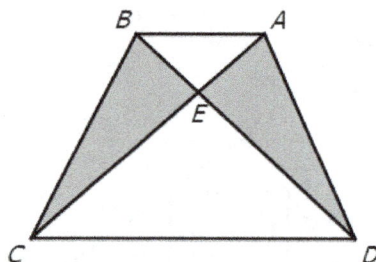

> **Tips:** *1. Think about the area properties found in a trapezoid. 2. Use the area method.*

Solution:

Because $\triangle AED$ and $\triangle BEC$ have equal areas, $ABCD$ is a trapezoid and $AB//CD$.

138

Knowing $AB//CD$, we get $AE/EC = AB/CD = 9/12 = 3/4$.

Therefore, $AE/AC = 3/(3+4) = 3/7$ and $AE = 14 * (3/7) = 6$.

The answer is (A).

Problem 73

Rectangle $ABCD$ has $AB = 4$ and $BC = 3$. Segment EF is constructed through B so that $EF \perp DB$, and A and C lie on DE and DF, respectively. What is EF ? (2009 AMC10)

(A) 9   (B) 10   (C) 125/12   (D) 103/9   (E) 12

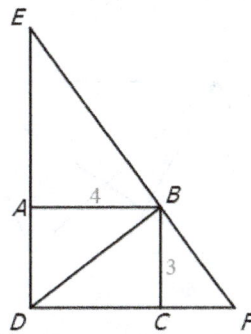

Tips: *Find pairs of similar right triangles and use their ratio properties.*

Solution 1:

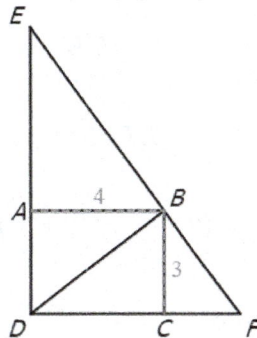

In the $Rt\triangle BDF$, BC is the altitude to the hypotenuse and we have $BC^2 = DC * CF$.

Solving the equation gives us $CF = 9/4$ and $DF = 25/4$. Similarly, $AB^2 = AE * AD$, and we have $ED = 3 + 16/3 = 25/3$. Thus $EF = 125/12$.

The answer is (C).

Solution 2:

139

We can clearly see that $\triangle ABE \sim \triangle CFB \sim \triangle CBD$ and $DB = 5$. Thus $EB/DB = AB/BC$, $EB = 5*4/3 = 20/3$.

Similarly, $BF/DB = BC/DC$, $BF = 5*3/4 = 15/4$. Therefore, $EF = EB + BF = 125/12$.

The answer is (C).

Problem 74

In parallelogram $ABCD$, points E and F trisect the segment AB. DF and DB intersect EC at G and H, respectively. If $EG : GH : HC = x : y : z$, where x, y and z are positive integers, what is the minimum possible value of $x + y + z$?

(A) 20   (B) 40   (C) 30   (D) 50   (E) 55

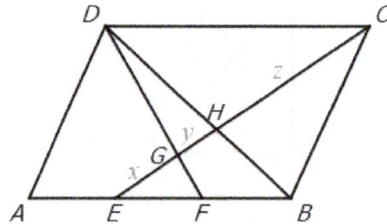

> **Tips:** *Calculate the ratio of the length of each individual segment in CE to the diagonal line using the properties of similar triangles.*

Solution:

There is an implicit parallel condition in the parallelogram ($AB//CD$ and $AD//BC$). The information $AB//CD$ leads to $\triangle GEF \sim \triangle GCD$ and $\triangle HEB \sim \triangle HCD$.

Because $\triangle GEF \sim \triangle GCD$, $EG/GC = EF/CD$. We get the equations $x/(y+z) = 1/3$ and $x/(x+y+z) = 1/4$.

Because $\triangle HEB \sim \triangle HCD$, $EH/HC = EB/CD$, thus $(x+y)/z = 2/3$ and $(x+y)/(x+y+z) = 2/5$.

Solving the equation set gives $y = (2/5 - 1/4)(x+y+z) = (3/20)*(x+y+z)$ and $z = (3/5)*(x+y+z)$.

Therefore, $x : y : z = 5 : 3 : 12$ and $x + y + z = 20$.

The answer is (A).

Problem 75

In parallelogram $ABCD$, points E and F are the midpoints of side AB and BC, respectively. AF intersects DE at G and BD at H. Find the area of quadrilateral $BHGE$ if the area of $ABCD$ is 60.

(A) 10   (B) 9   (C) 8   (D) 7   (E) 5

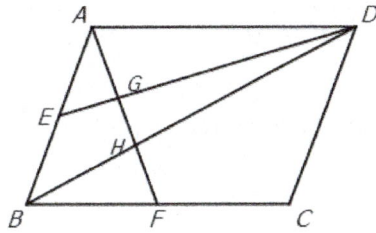

Tips: *Find the relationships among AG, GH and HF through similar triangle pairs.*

Solution 1:

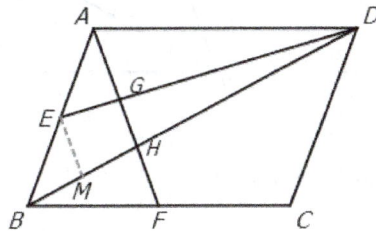

The key is to find the relationship between the lengths of AG, GH and HF. Let point M be on BH such that $EM//GH$.

It is easy to know $\Delta HAD \sim \Delta HBF$. Thus $AH/HF = HD/HB = AD/BF = 2$.

Therefore, $AH/AF = 2/3$, $HF/AF = 1/3$ and $BH/BD = 1/3$, $DH/BD = 2/3$.

Because $EM//AF$ and E is midpoint of AB, $EM = AH/2 = AF/3$ and $BM = MH = BD/6$.

Because $EM//AF$, $GH/EM = DH/DM = (2/3)/(1/6 + 2/3) = 4/5$.

Thus $GH = (1/2)*(4/5)*AH = (2/5)*AH = (4/15)*AF$ and $AG = (1 - 4/15 - 1/3)*AF = (2/5)*AF$.

Using the area method,

$$S_{\Delta DGH} = \frac{GH}{AH} * S_{\Delta DAH} = \frac{GH}{AH} * \frac{DH}{BD} * S_{\Delta ABD} = \frac{2}{5} * \frac{2}{3} * \frac{60}{2} = 8$$

The area of $BHGE$ is $60/4 - 8 = 7$.

The answer is (D).

Solution 2:

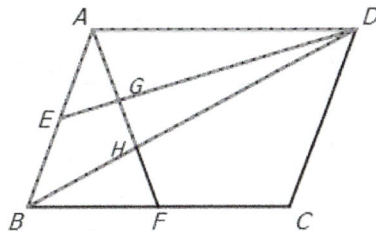

141

Similar to solution 1, we obtain $BH/HD = 1/2$. From here, we apply the mass point method in $\triangle ABD$.

The mass at points A, B, E and D, H are 1, 1, 2, 0.5 and 1.5, respectively. Thus along AH, $AG/GH = 1.5/1 = 3/2$.

Thus the area of $\triangle DGH$ is $30 * (DH/DB) * (GH/AH) = 30 * (2/3) * (2/5) = 8$ and the area of $BHGE$ is $60/4 - 8 = 7$.

The answer is (D).

Solution 3:

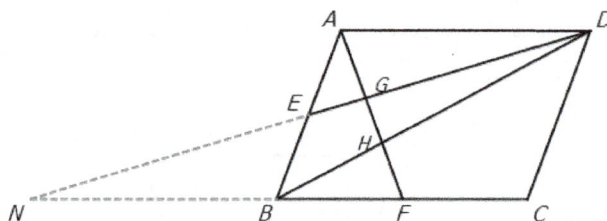

In this solution we use a slightly different way to find the ratios among AG, GH and HF.

Extend DE and CB and meet at point N. We use two pairs of similar triangles: $\triangle HAD \sim \triangle HBF$ and $\triangle GAD \sim \triangle GFN$.

Because $\triangle HAD \sim \triangle HBF$, $AH/HF = AD/BF = 2$ and $AH/AF = 2/3$.

Similarly, because $\triangle GAD \sim \triangle GFN$, $AG/GF = AD/NF = 1/1.5 = 2/3$ and $AG/AF = 2/5$.

Thus $GH/AF = 2/3 - 2/5 = 4/15$.

The rest of the process will be the same as in solutions 1 and 2.

The answer is (D).

Problem 76

In $\triangle ABC$, $\angle A = 2\angle B$, $AB = 4$ and $BC = 2\sqrt{3}$. Find the value of AC.

(A) 2   (B) 3   (C) 4   (D) 6   (E) 5

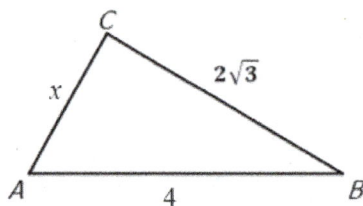

Tips: *1. Construct an isosceles triangle internally or externally at the presence of double angle. 2. Use the theorems related to angular bisectors (ratio and length).*

Solution 1:

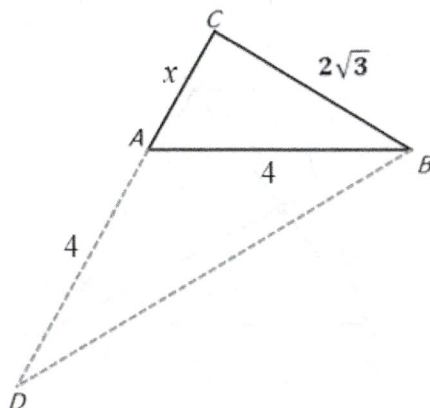

Construct an external isosceles triangle. Extend CA to D such that $AD = AB$ and connect BD. Because $AD = AB$ and $\angle CAB = 2\angle CBA$, $\angle ABD = \angle CBA$ (AB is the angular bisector of $\angle CBD$).

Let $y = BD$. Using the angular bisector length and ratio theorems, we have $BC/BD = AC/AD$ and $AB^2 = BC * BD - AC * AD$. Thus $2\sqrt{3}/y = x/4$ and $16 = y * 2\sqrt{3} - 4x$.

Simplifying the equation gives $x^2 + 4x - 12 = 0$ with the solutions $x = 2$ and $x = -6$ (ignore).

The answer is (A).

Solution 2:

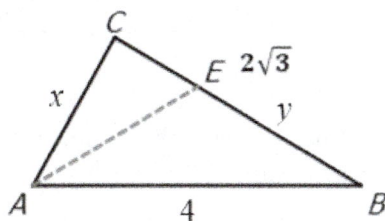

Construct an internal isosceles triangle based on the double angle information. Let point E be on side BC such that AE is the angular bisector of $\angle CAB$.

Because AE is angular bisector of $\angle CAB$ and $\angle CAB = 2\angle CBA$, $\angle EAB = \angle CBA$ and $AE = BE$.

Let BE be y and then $CE = 2\sqrt{3} - y$. From the angular bisector length and ratio theorems, we have the following two equations: $AC/AB = CE/EB$ and $AE^2 = AC * AB - CE * EB$. Thus $x/4 = (2\sqrt{3} - y)/y$ and $y^2 = 4x - y(2\sqrt{3} - y)$. Simplifying the equation gives $x^2 + 4x - 12 = 0$. Thus $x = 2$.

The answer is (A).

Problem 77

In trapezoid $ABCD$, $AB = 3CD$ and $AB//CD$. E is the midpoint of the diagonal AC. BE meets AD at F. Find the value of $AF : FD$.

(A) 5/3   (B) 3/2   (C) 10/7   (D) 8/5   (E) 12/5

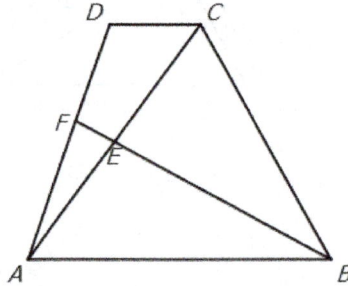

Tips: *1. Create pairs of similar triangles through extension of CD and BF. 2. Create parallel lines to transfer information from one edge to another.*

Solution 1:

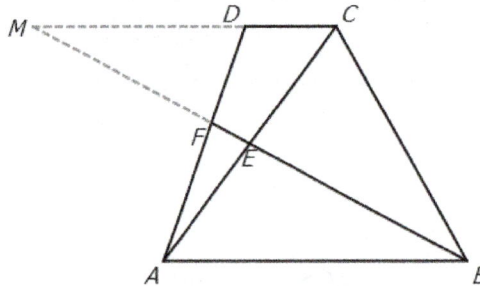

The idea here is to use parallel lines to construct multiple pairs of similar triangles. Extend segment BF and CD to meet at point M.

Because $CD//AB$, $\Delta EAB \sim \Delta ECM$ and $\Delta FAB \sim \Delta FDM$. Thus $AB/CM = AE/EC = 1$ and $AF/FD = AB/DM$

Therefore, $AF/FD = AB/DM = AB/(CM - CD) = 3/2$

The answer is (B).

Solution 2:

144

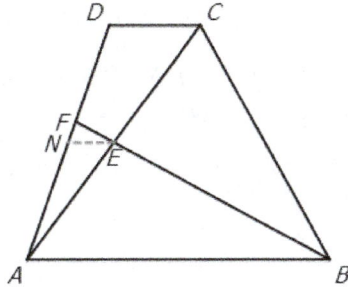

Let point N be on segment AF such that $NE//AB$.

Because $NE//AB//CD$ and $AE = EC$, $NE = CD/2 = AB/6$, $AN = ND$. $NE//AB$ leads to $FN/FA = NE/AB = 1/6$ and $FN/NA = 1/5$.

Thus $AF/FD = (NA + FN)/(NA - FN) = 6/4 = 3/2$.

The answer is (B).

Solution 3:

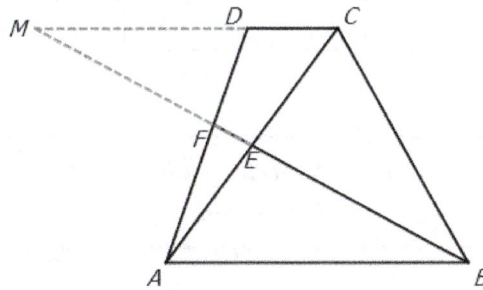

Similar to solution 1, extend segment BF and CD to meet at point M. $MC/MD = 3/2$. Applying **Menelaus' Theorem**, we have

$$\frac{MC}{MD} * \frac{FD}{AF} * \frac{AE}{EC} = 1$$

With $AE = EC$ and $MC/MD = 3/2$, we have $AF/FD = 3/2$.

The answer is (B).

Solution 4:

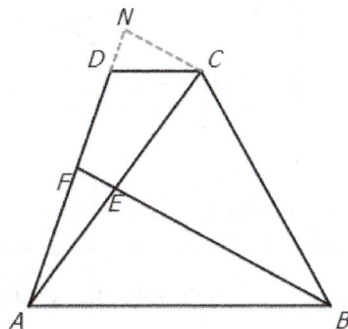

145

Let point N be on the extension of segment AD such that $CN//EF$.

Because $CN//EF$ and $AE = EC$, $AF = FN$. We also know $CN//EF$ and $CD//AB$, thus $\triangle EBC \sim \triangle EFD$ and $ND/AF = CD/AB = 1/3$.

Therefore, $AF/FD = AF/(AF - ND) = 1/(1 - 1/3) = 3/2$

The answer is (B).

Problem 78

In rectangle $ABCD$, point E and F are on side BC and CD, respectively. The areas of the $\triangle ABE$, $\triangle ECF$, and $\triangle FDA$ are 4, 3, and 5, respectively. Find the area of the $\triangle AEF$.

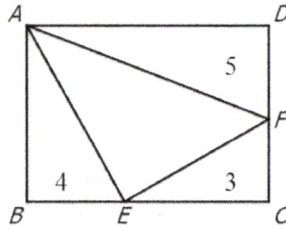

> **Tips:** *1. Represent each edge of the rectangle through the known areas. 2. Establish an equation set based on properties of rectangles.*

Solution 1:

Let $x = AB$ and $y = AD$. Then, $BE = 8/x$ and $DF = 10/y$. And $EC = y - BE = y - 8/x$, $FC = CD - DF = x - 10/y$.

Knowing that the area of triangle CEF, we have $EC * FC = 6$. Thus $(y - 8/x)(x - 10/y) = 6$. Simplifying the equation gives $xy + 80/xy = 24$.

Therefore, $xy = 20$ or $xy = 4$ (ignore) and the area of $\triangle AEF$ is $20 - 12 = 8$.

Solution 2:

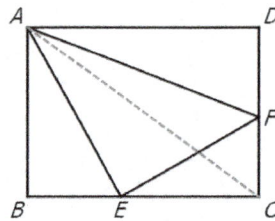

We will use the area method in this solution. Assume the area of $\triangle AEC$ and $\triangle AFC$ are x and y, respectively. $\triangle ACD$ and $\triangle ABC$ have equal area, thus $4 + x = 5 + y$.

In $\triangle AFC$ and $\triangle EFC$, we have $y/3 = AD/EC$. In $\triangle ABE$ and $\triangle ABC$, we have $BC/EC = (4 + x)/x$.

146

Because $AD = BC$, $y/3 = (4+x)/x$. Thus we have equation set $x = y+1$ and $xy = 12+3x$.

Solving the equation set gives $x = 6$ and $y = 5$. The area of $\triangle AEF$ is $11 - 3 = 8$.

Problem 79

In rectangle $ABCD$, points F and G lie on AB so that $AF = FG = GB$ and E is the midpoint of DC. Also, AC intersects EF at H and EG at J. The area of rectangle $ABCD$ is 70. Find the area of triangle EHJ.

(A) 5/2   (B) 35/12   (C) 3   (D) 7/2   (E) 35/8

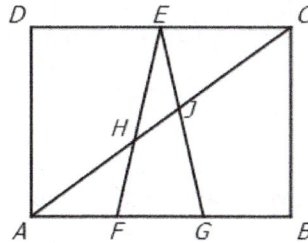

> **Tips:** *Use similar triangle pairs to find the ratio AH:HJ:JC and then use the area method.*

Solution 1:

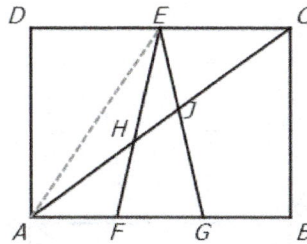

Connect AE. The area of $\triangle EAC$ is 17.5.

The next step is to find the ratio AH, HJ and JC through two pairs of similar triangles: $\triangle HAF \sim \triangle HCE$ and $\triangle JAG \sim \triangle JCE$. Because $\triangle HAF \sim \triangle HCE$, $AH/HC = AF/EC = 2/3$ and $AH/AC = 2/5$.

Similarly, because $\triangle JAG \sim \triangle JCE$, $AJ/JC = AG/EC = 4/3$. Thus $AJ/AC = 4/7$, $HJ/AC = 4/7 - 2/5 = 6/35$.

Using the area method, $S_{\triangle EHJ} = HJ/AC * S_{\triangle EAC} = 6/35 * 17.5 = 3$.

The answer is (C)

Solution 2:

147

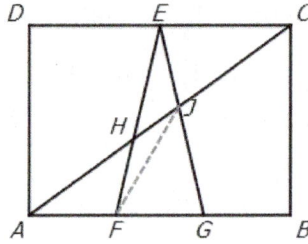

Similar to solution 1, we know the area of $\triangle EFG$ is $70/6$. It is straightforward to see that $\triangle HAF \sim \triangle HCE$ and $\triangle JAG \sim \triangle JCE$.

$\triangle HAF \sim \triangle HCE$ leads to $EH/HF = EC/AF = 3/2$ and $EH/EF = 3/5$. $\triangle JAG \sim \triangle JCE$ leads to $EJ/JG = EC/AG = 3/4$ and $EJ/EG = 3/7$.

Using the area method, $S_{\triangle EHJ} = (EJ/EG) * (EH/EF) * S_{\triangle EFG} = (3/7) * (3/5) * (70/6) = 3$.

The answer is (C).

Solution 3:

Connect AE and apply the mass point method on $\triangle EAG$. $\triangle JAG \sim \triangle JGE$ leads to $EJ/JG = EC/AG = 3/4$.

The mass on points A, F, G, E and J are 1, 2, 1, $4/3$ and $7/3$, respectively. Thus along AHJ, we have $HJ/AH = 3/7$. Thus $HJ/AJ = 3/10$ and $EJ/EG = 3/7$.

Using the area method, $S_{\triangle EHJ} = (EJ/EG) * (HJ/AJ) * S_{\triangle EAG} = (3/7) * (3/10) * (70/3) = 3$.

The answer is (C).

Problem 80

Let P be a point inside the rectangle $ABCD$. If $PA = 5$, $PD = 8$, and $PC = 12$, find PB.

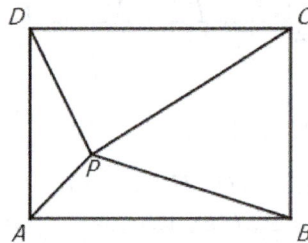

Tips: *Create right triangles and use the Pythagorean Theorem to discover a relationship between PA, PB, PC and PD.*

Solution:

148

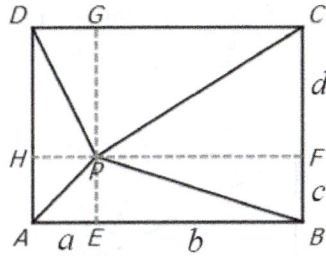

Construct HF and GE such that the rectangle is split into four small rectangles.

The length of each segment is labeled in the diagram. We have $PD^2 + PB^2 = a^2 + d^2 + c^2 + b^2 = PA^2 + PC^2$.

Therefore $PB^2 = PA^2 + PC^2 - PD^2 = 25 + 144 - 64 = 105$. Thus $PB = \sqrt{105}$.

Problem 81

$ABCD$ is a right trapezoid. E is the midpoint of BC. Prove: $AE = DE$.

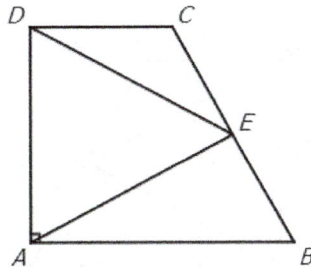

> **Tips:** *1. Prove the triangle is an isosceles triangle 2. Prove point E is the midpoint of the hypotenuse of a right triangle. 3. Find a pair of congruent triangles.*

Solution 1:

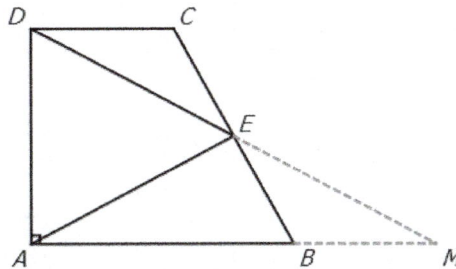

Extend AB and DE to meet at point M.

Because $CD//AB$ and $CE = EB$, $DE = EM$

149

In $Rt\triangle DAM$, AE is the median of the hypotenuse. Thus $AE = DE$

Solution 2:

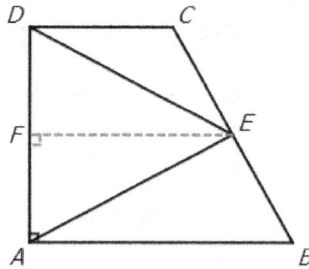

Let point F be on AD such that $EF // CD$.

$\because EF // CD // AB$, $AB \perp AD$ and $CE = EB$

$\therefore DF = FA$ and $EF \perp AD$

$\therefore FE$ is the perpendicular bisector of segment AD.

$\therefore AE = DE$

Solution 3:

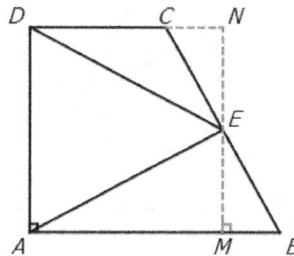

Extend DC to N such that $NM \perp AB$ at point M.

Therefore $\angle N = \angle EMB = 90°$. With $CE = EB$, we have $\triangle ECN \cong \triangle EBM$.

$\therefore EN = EM$

$\because DN = AM$, $EN = EM$ and $\angle N = \angle EMA = 90°$

$\therefore \triangle EDN \cong \triangle EAM$

$\therefore AE = DE$

Problem 82

In trapezoid $ABCD$, $AB // DC$, $BC = CD = 7$ cm, $AD = 8$ cm, and $BD \perp AD$. Find the length of AB.

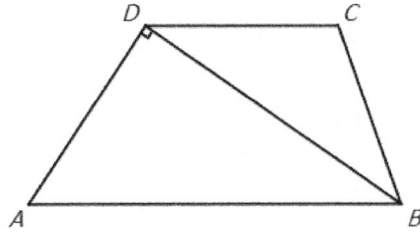

> **Tips:** *Since this geometry-rich problem involves a right triangle, an isosceles triangle, an angular bisector, and parallel lines, there are multiple ways to solve this problem.*

Solution 1:

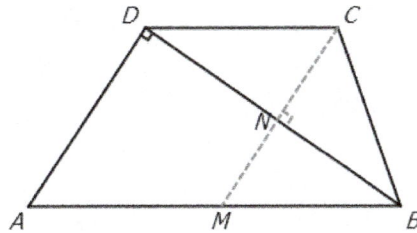

This problem involves nearly all important topics in geometry such as right triangles, parallel lines, isosceles triangles, and angular bisectors.

Let point N be the midpoint of BD and extend CN to meet AB at point M.

$\because BC = CD, DN = BN$

$\therefore CN \perp BN$ and $\angle CDB = \angle CBD$

$\because CD // AB$

$\therefore \angle DBA = \angle CDB = \angle CBD$ (BD is the angular bisector of $\angle ABC$)

$\because \angle DBA = \angle CBD$ and $CN \perp BN$

$\therefore BM = BC = 7$

$\because CD // AB$ and $CM // AD$

$\therefore AMCD$ is a parallelogram.

$\therefore AM = CD = 7$.

Therefore, $AB = AM + BM = 7 + 7 = 14$

Solution 2:

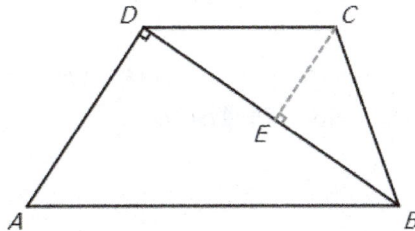

Let point E be the midpoint of BD and connect CE.

$\because BC = CD$, $DE = BE$

$\therefore CE \perp BD$ and $\angle CDB = \angle CBD$

$\because CD // AB$

$\therefore \angle DBA = \angle CDB$

$\because \angle DBA = \angle CDB$ and $\angle DEC = \angle BDA = 90°$

$\therefore \triangle CDE \sim \triangle ABD$

$\therefore AB/CD = BD/ED = 2$

Therefore, $AB = 2CD = 14$.

Solution 3:

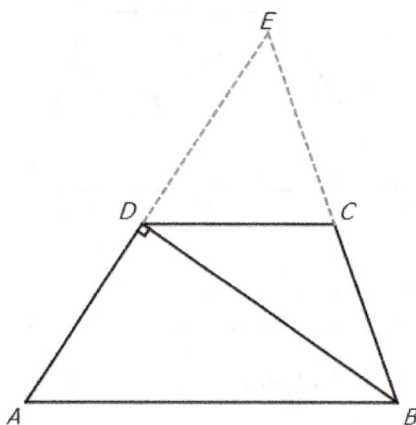

Extend BC and AD to meet at point E.

In the right triangle BDE, $CD = CB$ leads to $CE = CB = 7$ (CD is the median of BE in a right triangle).

$\because CD // AB$ and $\angle CDB = \angle CBD$

$\therefore \angle ABD = \angle CBD$

$\because BD // AD$

$\therefore \triangle ABC$ is a isosceles triangle

Therefore, $AB = BE = CE + CD = 14$.

Problem 83

The trapezoid $ABCD$ has $AD // BC$. $\angle A = 90°$. $AB = AD = 2CD$. $PD = 1$. $PA = 2$. $PB = 3$. Find the area of the trapezoid $ABCD$.

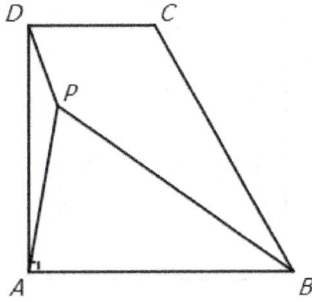

Tips: *1. To solve the problem, we need to know the length of one side only. 2. Rotate a triangle to construct a new triangle that contains all the information.*

Solution 1:

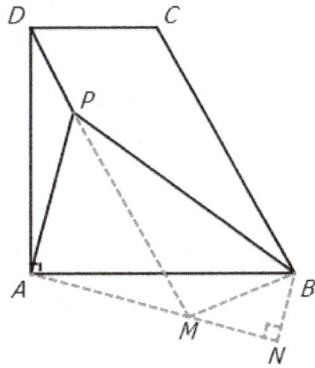

To solve this problem, we need to know the length of one side only, e.g., AD.

Knowing $AB = AD$ and $\angle A = 90°$, we can make a rotational transformation. Rotate triangle DPA 90 degrees clockwise around point A to create BMA. Thus we have $AM = AP = 2$, $PM = 2\sqrt{2}$, and $BM - PD = 1$, $\angle PMA = 45°$ (we also know $\angle APD = \angle AMB = 135°$).

$\because$ In $\triangle PBM$, $PB^2 = PM^2 + BM^2 = 9$

$\therefore BM \perp PM$

Extend AM to point N such that $BN \perp AM$.

$\because BN \perp AM$, $\angle BMN = \angle PMA = 45°$

$\therefore MN = BN = BM/\sqrt{2} = \sqrt{2}/2$

Therefore, $AN = 2 + \sqrt{2}/2$ and $AB^2 = 5 + 2\sqrt{2}$.

And the area of $ABCD$ is

$$(AB + CD) * AD/2 = (3/4) * AB^2 = \frac{15 + 6\sqrt{2}}{4}$$

Solution 2:

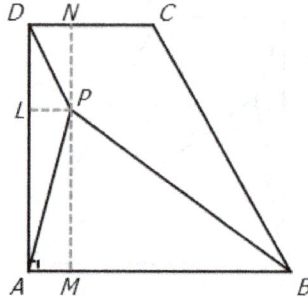

Construct altitude PM and PL on AB and AD, respectively. Assume $DL = x$, $PL = y$ and $AD = z$.

Using the **Pythagorean Theorem**, we have $PD^2 = DL^2 + PL^2$, $PA^2 = PL^2 + LA^2$, and $PB^2 = MB^2 + LA^2$.

Thus we have the following three equations: $x^2 + y^2 = 1$, $(z - x)^2 + y^2 = 4$, and $(z - y)^2 + (z - x)^2 = 9$

It may take some effort to solve the quadratic equation set. Nevertheless, we only need to find z^2. Using the first and second equations, as well as the first and third equations, we obtain

$$x = \frac{z^2 - 3}{2z}$$

and

$$y = \frac{z^2 - 5}{2z}$$

Substituting x and y into the first equation, we get $z^2 = 5 + 2\sqrt{2}$. Therefore, the area of $ABCD$ is

$$(AB + CD) * (AD/2) = (3/4) * z^2 = \frac{15 + 6\sqrt{2}}{4}$$

Solution 3:

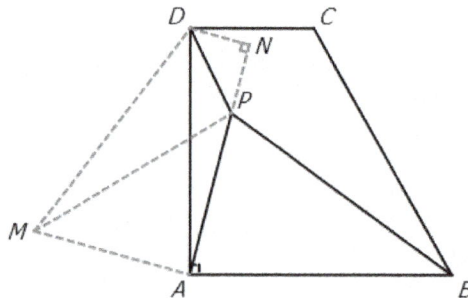

The approach is similar to that in solution 1, except that we rotate triangle ABP counter-clockwise around point A to create ADM and then connect MP. We can prove a right triangle MPD and $\angle APD = 135°$. The rest of the process is similar to solution 1.

Solution 4:

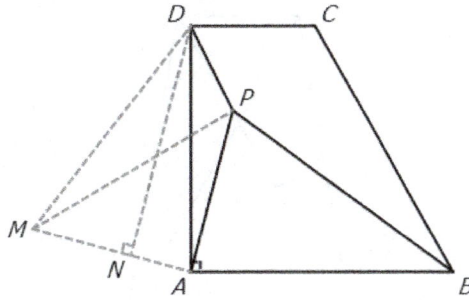

This method is similar to solution 3, but here we construct the altitude on MA for the triangle DMA. To calculate the length of DN or MN, we need to know $\sin \angle DMA$.

From solution # 3, we know $\sin \angle DMP = 1/3$ and $\sin \angle PMA = 1/\sqrt{2}$. Thus $\sin \angle DMA = \frac{\sqrt{2}+4}{6}$ and $DN = DM \sin \angle DMA$ and $NA = 2 - DM \cos \angle DMA$.

Therefore, $AD^2 = 5 + 2\sqrt{2}$ and the area of $ABCD$ is

$$(AB + CD) * AD/2 = (3/4) * AB^2 = \frac{15 + 6\sqrt{2}}{4}$$

Problem 84

$ABCD$ is an isosceles trapezoid with $AD = 10$. The diagonal AC and BD meet at G. $\angle AGB = 60°$. If E is the midpoint of GB and F is the midpoint of AB, what is the length of EF?

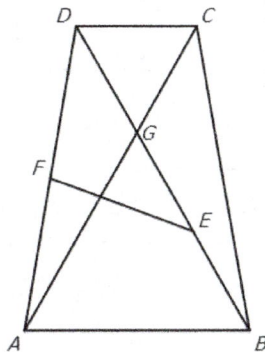

> **Tips:** *Connect AE and use the median property of an equilateral triangle.*

Solution:

155

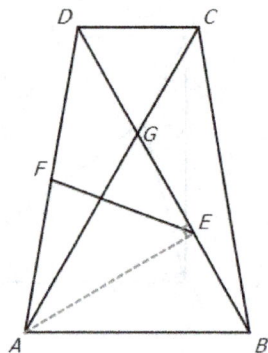

Connect AE. In the equilateral triangle ABG, E is the midpoint of BG, thus $AE \perp BG$.

Because the ΔHCD is a right triangle and F is the midpoint of AD, $EF = AD/2 = 10/2 = 5$.

Problem 85

In a square $ABCD$, $AB = 2$ and there are four quarter circles of radius 1 at each corner of the square. What is the area of the shaded area in the diagram?

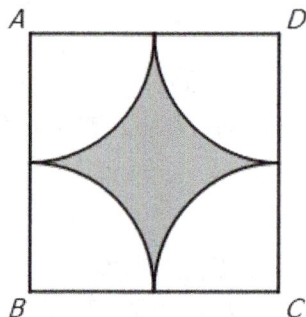

> **Tips:** *The total area of the quarter circles is the area of a full circle.*

Solution:

The four quarter circles form a full circle with a radius of 1 and an area of π. Therefore, the shaded area is $4 - \pi$.

Problem 86

Each of the small circles has radius 3. The innermost circle is tangent to the six circles that surround it, and each of those circles is tangent to the large circle and to its small-circle neighbors. What is the ratio of the area of the shaded region to the area of the unshaded regions?

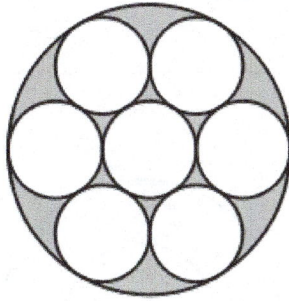

Tips: *Find the ratios between the diameters of the small and large circles.*

Solution:

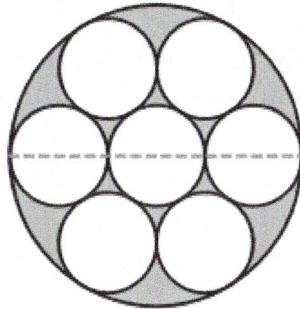

From the red dashed line we know the ratio of the diameters between small and large circle is 1 : 3, so the area ratio of one small circle and the big circle is 1 : 9.

We have 7 small circles, which leads to the area ratio of all small circles to the big circle to be 7 : 9. Therefore, the ratio of the shaded area to the unshaded area is 2 : 7.

Problem 87

The two circles have the same center C. Chord AD of the outer circle is tangent to the inner circle. If $AD = 24$, what is the shaded area.

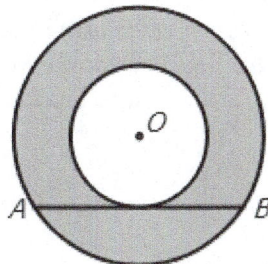

Tips: *Connect OA and tangent point to form a right triangle.*

Solution:

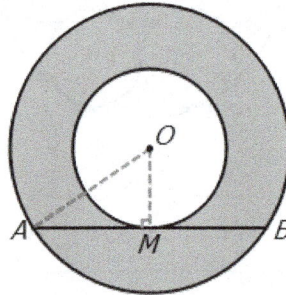

Connect OA and OM, where M is the tangent point.

We know OM is the perpendicular bisector of AB, thus $OM \perp AB$ and $AM = MB = 24/2 = 12$. The **Pythagorean Theorem** gives $AM^2 = OA^2 - OM^2 = 144$. Therefore, the shaded area is $\pi(OA^2 - OM^2) = 144\pi$.

Problem 88

The inscribed circle I of $\triangle ABC$ divides segment BC into two segments of lengths 6 and 8 by the tangent point. If the radius of the circle is 4, find the length of the longest side of the triangle?

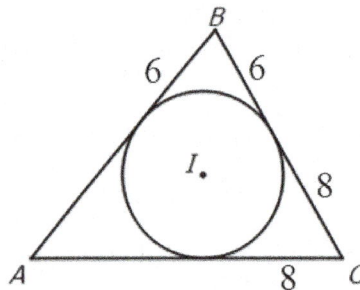

Tips: *Establish an equation through calculating the area of the triangle in two separate ways. (Heron's Formula and inscribed circle method).*

Solution:

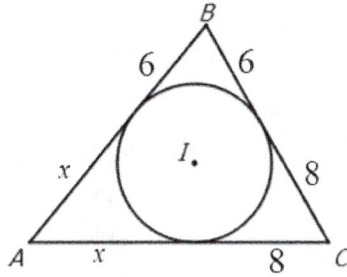

Use **Heron's Formula** to calculate the triangle area.

As show in the diagram, $s = 6 + 8 + x$, **Heron's Formula** gives area

$$\sqrt{s(s-6)(s-8)(s-x)} = rs$$

Then we have $\sqrt{(14+x)(6+x)(8+x)14} = 4(14+x)$.

Solving this equation gives $x = 7$, thus the longest side of the triangle is 15.

Problem 89

A circle passes through the three vertices of an isosceles triangle that has two sides of length 3 and a base of length 2. What is the area of this circle? (2007 AMC10A)

(A) 2π (B) $\frac{5}{2}\pi$ (C) $\frac{81}{32}\pi$ (D) 3π (E) $\frac{7}{2}\pi$

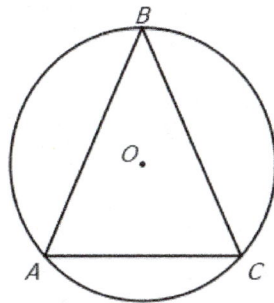

> **Tips:** *Connect the circumcenter to A and draw the perpendicular bisector to AC.*

Solution 1:

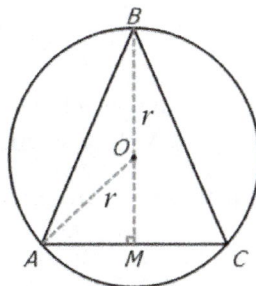

Given an isosceles triangle and its circumcicle, it is intuitive to construct the perpendicular bisector of triangle's base and connect the circumcenter to its vertices.

In the diagram, M is the midpoint of AC. Applying the **Pythagorean theorem** to $Rt\triangle OAM$ and $Rt\triangle ABM$ gives $r^2 = OM^2 + 1$ and $3^2 = BM^2 + 1$.

With $BM = r + OM$, we have $r + \sqrt{r^2 - 1} = 8$. The way to solve this equation is to transform it to $r - 8 = \sqrt{r^2 - 1}$ and take square on both sides, which leads to $r = 9/(4\sqrt{2})$.

Therefore, the the area of the circle is $\frac{81}{32}\pi$.

The answer is (A).

Solution 2:

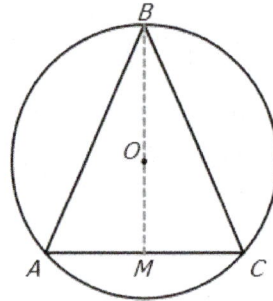

The law of sines can be used to find the radius of the circumcircle.

In right $\triangle ABM$, we have $\sin \angle A = 2\sqrt{2}/3$. Applying the law of sines, we have $2r = 3\sin \angle A = 9/(2\sqrt{2})$, $r = 3\sin \angle A = 9/(4\sqrt{2})$.

Then, the the area of the circle is $\frac{81}{32}\pi$.

The answer is (A).

Solution 3:

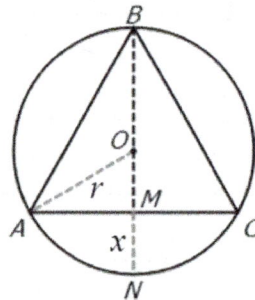

Extend the perpendicular bisector BM and meet the circle at point N. Then we can use the **Power of a Point Theorem** and the **Pythagorean Theorem** to solve this problem.

Let $x = MN$. The **Power of a Point Theorem** gives $x(2r - x) = 1$. In $Rt\triangle ABM$, we have $(2r - x)^2 = 8$. Solving the equation set gives $x = 1/\sqrt{8}$ and $r = 9/(4\sqrt{2})$.

Therefore, the area of the circle is $\frac{81}{32}\pi$.

The answer is (A).

Problem 90

The number of inches in the perimeter of an equilateral triangle equals the number of square inches in the area of its circumscribed circle. What is the radius, in inches, of the circle? (2003 AMC10A)

(A) $3\sqrt{2}/\pi$　　(B) $3\sqrt{3}/\pi$　　(C) $\sqrt{3}$　　(D) $6/\pi$　　(E) $\sqrt{3}\pi$

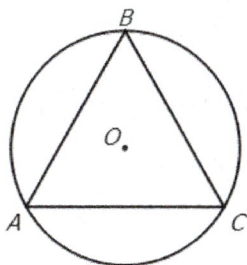

Tips: *Connect circumcenter to A and use perpendicular bisector on AC to for right triangles.*

Solution 1:

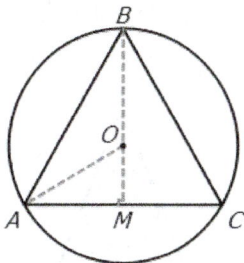

Let point M be the midpoint of AC and connect BM and OA. The circumcenter O is on BM.

In the right triangle OAM, we have $AM = \frac{\sqrt{3}}{2}r$.

The perimeter of the triangle is $3\sqrt{3}r = \pi r^2$.

Therefore $r = 3\sqrt{3}/\pi$.

The answer is (B).

Solution 2:

161

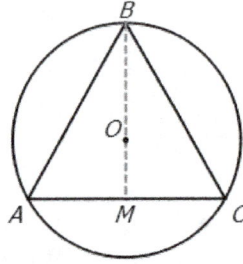

We can use the law of sines to find the radius of the circumcircle.

Let the edge-length of the equilateral triangle be a.

In triangle ABC, we have $\sin \angle A = \sqrt{3}/2$.

Applying the law of sines, we have $2r = a/\sin \angle A = a/(\sqrt{3}/2)$.

We also have $3a = \pi r^2$. Therefore $r = 3\sqrt{3}/\pi$.

The answer is (B).

Problem 91

A circle of radius 1 is tangent to a circle of radius 2. The sides of $\triangle ABC$ are tangent to the circles as shown, and sides AB and AC are congruent. What is the area of $\triangle ABC$? (2006 AMC10A)

(A) $35/2$ (B) $15\sqrt{2}$ (C) $64/3$ (D) $16\sqrt{2}$ (E) 24

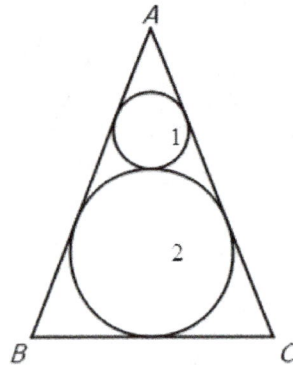

> **Tips:** *Draw the perpendicular bisector to BC and connect tangent points to centers of the circles to create similar triangles.*

Solution 1:

162

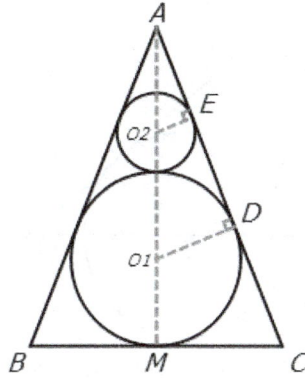

Draw AM, where M is the midpoint of side BC. Points D and E are the tangent points on side AC. Then we have $O_1D \perp AC$, $O_2E \perp AC$ and $AM \perp BC$. Therefore, $\triangle AO_2E \sim \triangle AO_1D \sim \triangle AMC$. Then we have $AO_2/AO_1 = O_2E/O_1D = 1/2$ and $AO_2 = O_1O_2 = 3$ and $AM = 8$.

Applying the **Pythagorean Theorem** in $\triangle AO_2E$ gives $AE = 2\sqrt{2}$. Therefore, $MC = (AM/AE) * O_2E = 2\sqrt{2}$. The area of triangle is $8 * 2\sqrt{2} = 16\sqrt{2}$.

The answer is (D).

Solution 2:

Similar to solution 1, we have $AM = 8$ and $AE = 2\sqrt{2}$. Thus $\tan \angle MAC = (1/2) * \sqrt{2}$ and $MC = AM * \tan \angle MAC = 2\sqrt{2}$. Therefore, the area of triangle is $8 * 2\sqrt{2} = 16\sqrt{2}$.

The answer is (D).

Problem 92

ABC is a triangle with $BC = 28$, $AC = 25$, and $AB = 17$. Circle O has the center on BC. D and E are the tangent points. Find the distance from O to D.

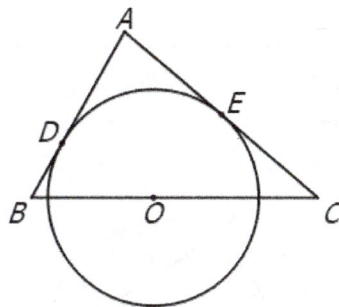

Tips: *Establish an equation through calculating the area of the triangle in two separate ways. (Heron's Formula and inscribed circle method)*

163

Solution:

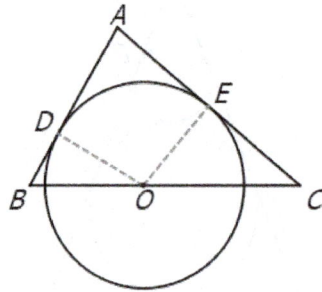

For problems involving tangent lines, it is helpful to connect tangent points to center of circle, e.g., OD and OE. The length of each sides of the triangle is known.

Using **Heron's Formula**, where $s = (a + b + c)/2 = 35$, we obtain that the triangle's area is $\sqrt{s(s - a)(s - b)(s - c)} = 210$.

On the other hand, the area of the triangle is equal to $(AB + AC) * r/2 = 21r$. Therefore, $OD = r = 10$.

Problem 93

A circle is circumscribed about a triangle with sides 8, 15, and 17, which divides the interior of the circle into four regions. What is the total area of the shaded regions?

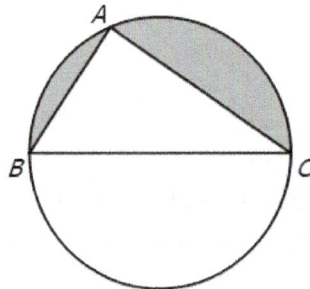

Tips: *Recognize triangle ABC is a right triangle.*

Solution:

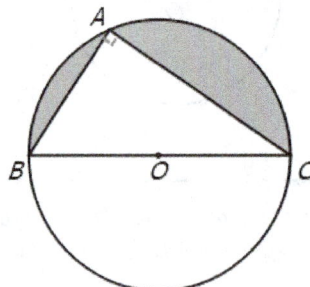

164

In triangle ABC, because $8^2 + 15^2 = 17^2$, the converse of the **Pythagorean Theorem** gives $AC \perp AB$.

Therefore, BC is the diameter of the circle and the shaded area is $\frac{17^2}{8}\pi - 60 = \frac{289}{8}\pi - 60$.

Problem 94

Let AB be a diameter of a circle and C be a point on AB with $2AC = BC$. Let D and E be points on the circle such that $DC \perp AB$ and DE is a second diameter. What is the ratio of the area of $\triangle DCE$ to the area of $\triangle ABD$? (2005 AMC10A)

(A) 1/6   (B) 1/4   (C) 1/3   (D) 1/2   (E) 2/3

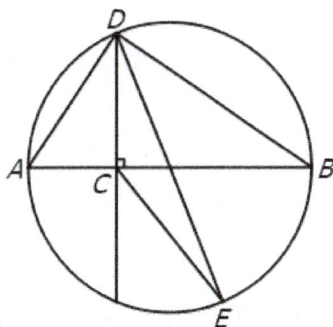

> **Tips:** *Find the altitude of triangle CDE and its relationship to the diameter.*

Solution 1:

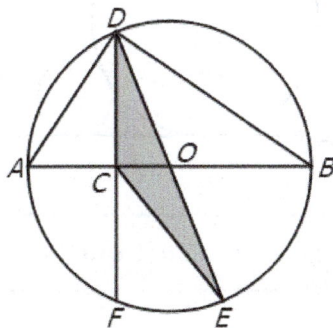

Let $r = OD = OE$, then $S_{\triangle CDE} = 2S_{\triangle COD}$.

We are given $CB = 2AC$, which leads to $(r + OC) = 2(r - OC)$, thus $OC/r = 1/3$ and $OC/AB = 1/6$.

Therefore $S_{\triangle COD} = (OC/AB) * S_{\triangle ABD} = OC/(2r) * S_{\triangle ABD} = (1/6) * S_{\triangle ABD}$ and $S_{\triangle CDE} = 2S_{\triangle COD} = (1/3) * S_{\triangle ABD}$.

The answer is (C).

165

Solution 2:

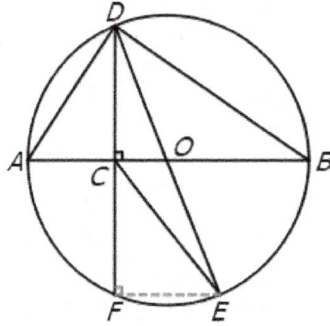

Connect EF, and it is clear that $EF \perp DF$. Similar to solution 1, we have $OC = (1/6) * AB$.

Because $EF \perp DF$ and $OC \perp CD$, $OC // EF$ and $OD = OE$. Therefore, $EF = 2OC = (1/3) * AB$. Then $S_{\triangle CDE} = EF * CD = (1/3)AB * CD = (1/3) * S_{\triangle ABD}$.

The answer is (C).

Problem 95

The two tangent circles with the radius of 3 and 1, respectively, have an external common tangent line AB. What is the shaded area?

(A) $4\sqrt{3} - \frac{11}{6}\pi$ (B) $4\sqrt{3}$ (C) $\frac{11}{6}\pi$ (D) 2π (E) $4\sqrt{3} - \frac{7}{6}\pi$

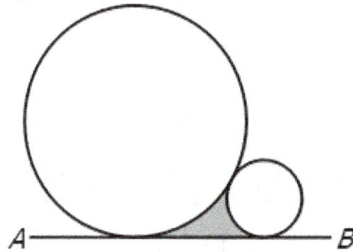

> **Tips:** *Connect the centers of the two circles and their respective tangent points.*

Solution:

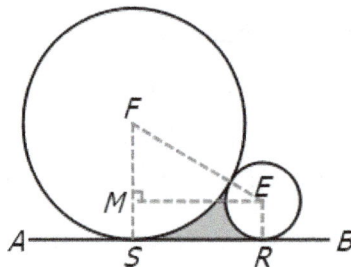

166

As shown in the diagram, points E and F are the centers of the circles and points S and R are the tangent points.

Connect EF, FS and ER. Let M be a point on FS such that $EM \perp FS$.

In the $Rt\triangle EFM$, $EF = 4$, $FM = 2$ and $EM = 2\sqrt{3}$. So $\angle F = 60°$ and $\angle FEM = 30°$.

The area of $EFSR$ (trapezoid) is $(1 + 3) * 2\sqrt{3}/2 = 4\sqrt{3}$.

The two sectors of the circle have an area of $9/6\pi$ and $\pi/3$.

Therefore the shaded area is $4\sqrt{3} - \frac{11}{6}\pi$.

The answer is (A).

Problem 96

Circles with centers A and B have radii 3 and 8, respectively. A common internal tangent intersects the circles at C and D, respectively. Lines AB and CD intersect at E, and $AE = 5$. What is CD? (2006 AMC10A)

(A) 13   (B) 44/3   (C) $\sqrt{221}$ (D) $\sqrt{255}$ (E) 55/3

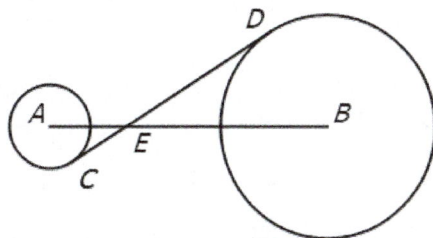

Tips: *Construct a right triangle and a rectangle by connecting the centers of the circles to the tangent points.*

Solution 1:

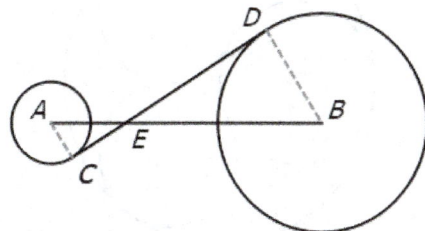

Connect AC and BD. Both AC and BD are perpendicular to CD.

Applying the **Pythagorean theorem** on $\triangle ACE$ gives $CE = 4$.

Because $\triangle ACE \sim \triangle BDE$, $DE/CE = BD/AC$, thus $DE = 8/34 = 32/3$. And $CD = 4 + 32/3 = 44/3$.

The answer is (B).

Solution 2:

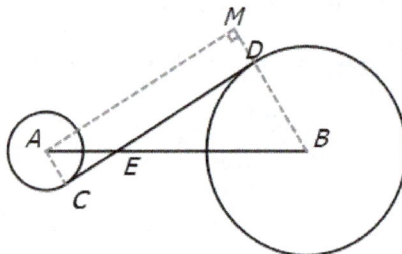

A useful technique for problems involving internal tangents is to construct a right triangle as shown in the diagram. $ACDM$ is a rectangle and ABM is a right triangle.

Because $\triangle ACE \sim \triangle BDE$, $EB/EA = BD/AC$, thus $BE = 8/35 = 40/3$ and $AB = 55/3$. Therefore, $BM = 8 + 3 = 11$.

Applying the **Pythagorean Theorem** in triangle ABM gives $AM = \sqrt{AB^2 - BM^2} = 44/3 = CD$.

The answer is (B).

Problem 97

Circles A, B, and C are externally tangent to each other and internally tangent to circle D. Circles B and C are congruent. Circle A has radius 1 and passes through the center of circle D. What is the radius of circle B? (2004 AMC10A)

(A) 2/3   (B) $\sqrt{3}/2$ (C) 7/8   (D) 8/9   (E) $(1+\sqrt{3})/3$

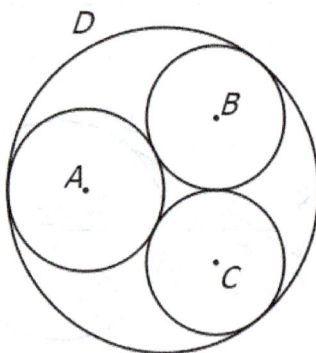

Tips: *Construct triangles by connecting the centers of the circles and the tangent points.*

168

Solution:

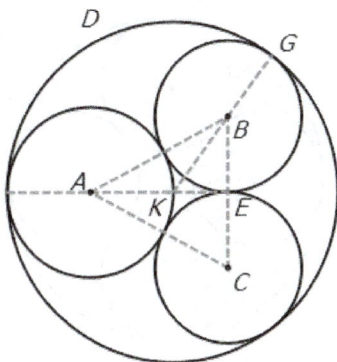

Let points A, B and C be the centers of the three small circles and G be the tangent point to the large circle.

Because circles B and C are congruent and the center of circle D is on circle A, the center of circle D, denoted by K, is on the perpendicular bisector (AE) of BC. Points K, B and G are colinear. $AK = 1$ and $KG = 2$.

Let $r = BG$. Applying the **Pythagorean Theorem** to the $Rt\triangle ABE$ and $Rt\triangle KBE$ gives $AE^2 = (1+r)^2 - r^2 = 2r + 1$ and $KE^2 = (2-r)^2 - r^2 = 4 - 4r$.

Because $AE = 1 + KE$, $\sqrt{2r+1} = 1 + \sqrt{4-4r}$. Solving this equation gives $r = 8/9$ or $r = 0$ (ignore).

The answer is (D).

Problem 98

Four congruent circles of radius 2 are circumscribed by a larger circle. Find the area of the shaded regions.

(A) $(9 + 2\sqrt{2})\pi - 16$ (B) $(9 + 2\sqrt{2})\pi - 4$ (C) $9\pi - 16$ (D) $2\sqrt{2}\pi - 16$

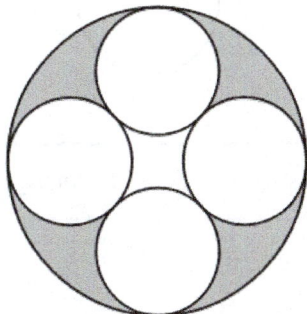

Tips: *Construct a square through the centers of the four small circles.*

169

Solution:

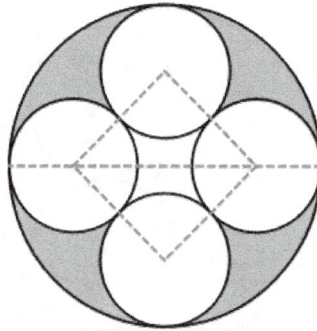

Connect the center of neighboring circles to form a square with side-length of 4.

The radii of the big circle is $2(1 + \sqrt{2})$. The shaded area is equal to the difference between the areas of the big circle and the the area of the four 3-quarter circles plus the square, which is $(12 + 8\sqrt{2})\pi - 3 * 4\pi - 16 = 8\sqrt{2}\pi - 16$.

The answer is (D).

Problem 99

A circle of radius 1 is surrounded by 4 circles of radius r. What is the value of r ?

(A) $\sqrt{2}$ (B) $1 + \sqrt{2}$ (C) $\sqrt{6}$ (D) 3   (E) $2 + \sqrt{2}$

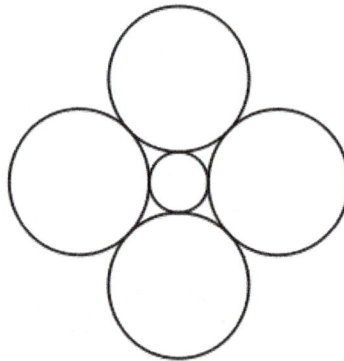

Tips: *1. Connect the centers of the circles to form a square and focus on the diagonals.*

Solution 1:

170

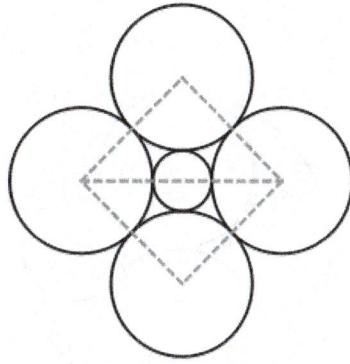

Connect the centers of the four circles to form a square.

The diagonal of the square and two sides form an isosceles right triangle with a side-length of $2r$ and hypotenuse-length of $2r + 2$. We get the following equation: $2r + 2 = 2r\sqrt{2}$.

Solving the equation gives $r = 1 + \sqrt{2}$.

The answer is (B).

Solution 2:

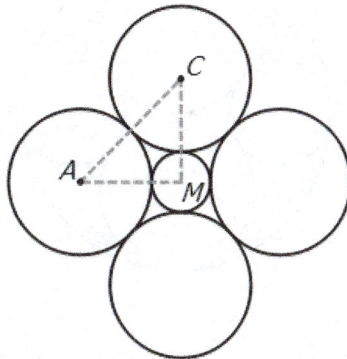

Connect the centers of the small circle and two neighboring circles to form an isosceles right triangle ACM.

We know $AC = 2r$, $AM = 1 + r$, and $AC = \sqrt{2}AM$, which leads to the equation $2r = \sqrt{2}(1 + r)$.

Solving the equation gives $r = 1 + \sqrt{2}$.

The answer is (B).

Problem 100

Two congruent circles are tangent to each other and inscribed by a larger circle. The other two smaller congruent circles are tangent neighboring circles externally. What is the ratio of the areas of the shaded region to the largest circle?

171

(A) 23/36   (B) 13/36   (C) 13/18   (D) 5/18   (E) 1/3

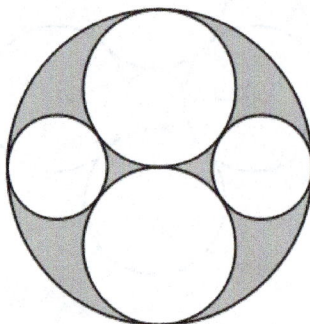

> **Tips:** *Connect centers of circles to form a right triangle and connect tangent points to circle center.*

Solution:

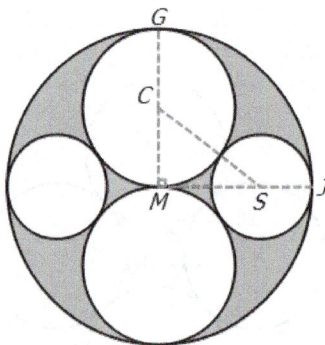

Let points M, C and S be the centers of the large, medium and small circles, respectively.

CMS is a right triangle. Let $x = CM$, $y = SJ$. By the **Pythagorean Theorem**, $CS^2 = CM^2 + MS^2$. So $(x + y)^2 = x^2 + (2x - y)^2$. Solving the equation gives us $y = \frac{2}{3}x$.

The area of the largest, medium and small circles is $4x^2\pi$, $x^2\pi$ and $(4/9)x^2\pi$, respectively, thus the shaded area is $(10/9)x^2\pi$.

Therefore, the ratio of the areas of the shaded region to the largest circle is 5/18.

The answer is (D).

Problem 101

A circle of radius 1 is surrounded by 4 circles of radius r as shown. What is the value of r? (2007 AMC10B)

(A) $\sqrt{2}$ (B) $1 + \sqrt{2}$ (C) $\sqrt{6}$ (D) 3   (E) $2 + \sqrt{2}$

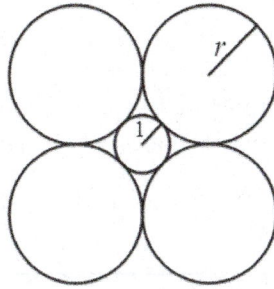

Tips: *Connect centers of circles to form an isosceles right triangle.*

Solution 1:

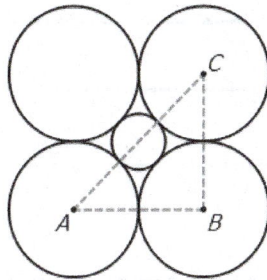

As shown in the diagram, connect three centers, A, B and C of the large circles. AC passes through the center of the small circle.

Triangle ABD is an isosceles right triangle with $AB = BC = 2r$ and $AC = 2r + 2$. By the **Pythagorean Theorem**, $AC = \sqrt{2} * AB$. Thus $2r + 2 = 2\sqrt{2}r$. Therefore, $r = \sqrt{2} + 1$.

The answer is (B).

Solution 2:

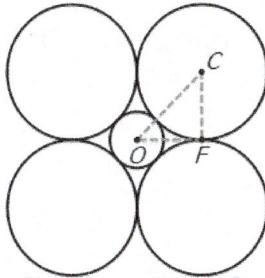

Connect the center of the small circle (O) with the center of one of the large circles (C) and its tangent point (F).

Triangle OCF is an isosceles right triangle with $CF = OF = r$ and $OC = 1 + r$. By the **Pythagorean Theorem**, $OC = \sqrt{2}CF$. Thus $r + 1 = \sqrt{2}r$. Therefore, $r = \sqrt{2} + 1$.

The answer is (B).

Problem 102

Four circles of radius 1 are each tangent to two sides of a square and externally tangent to a circle of radius 2, as shown. What is the area of the square? (2007 AMC10A)

(A) 32   (B) $22 + 12\sqrt{2}$ (C) $16 + 16\sqrt{3}$ (D) 48   (E) $36 + 16\sqrt{2}$

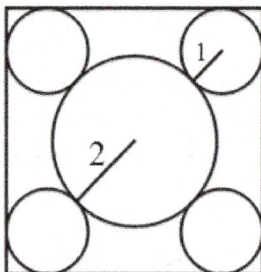

> **Tips:** *Connect the centers of circles to form an isosceles right triangle.*

Solution 1:

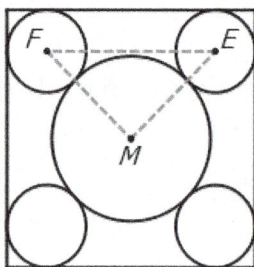

Connect the centers of two small circles (E and F) with the center of the large circle (M).

Triangle MEF is an isosceles right triangle with $ME = MF = 3$ and $FE = \sqrt{2}ME = 3\sqrt{2}$. The edge-length of the square is $FE + 2 = 2 + 3\sqrt{2}$. Therefore, the area of square is $22 + 12\sqrt{2}$.

The answer is (B).

Solution 2:

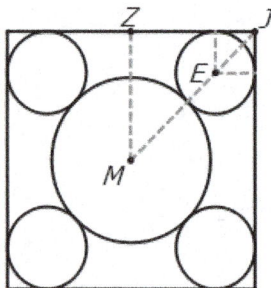

Connect the center of the circle with a radius of 2 (M) to the vertex of the square (J) and the midpoint of the squares side (Z).

MZJ is an isosceles right triangle with $MJ = ME + EJ = 1 + 2 + \sqrt{2} = \sqrt{2}ZJ$. Thus the edge-length of the square is $2 + 3\sqrt{2}$. Therefore, the area of square is $22 + 12\sqrt{2}$.

The answer is (B).

Problem 103

Two circles of radii of 4 and 6 are externally tangent and are circumscribed by a larger circle. Find the ratio of the area of the shaded region to the area of the unshaded region?

(A) 3/19   (B) 6/19   (C) 7/19   (D) 8/17   (D) 5/7

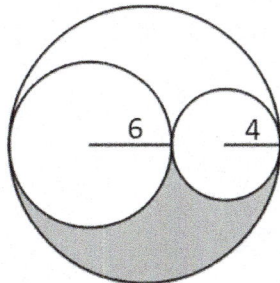

> **Tips:** *The shaded area is half of the difference between areas of external circle and two internal circles.*

Solution 1:

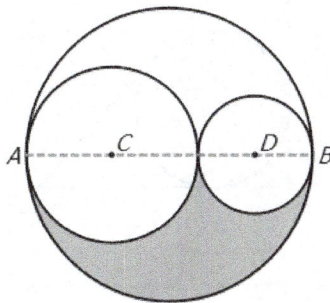

The shaded area is the difference between the area of the large half-circle and two small half-circles.

Given $AB = 20$, $DB = 4$ and $AC = 6$, the areas of the three half-circles are 50π, 8π and 18π, respectively. Thus the area of the shaded region is 24π and the area of unshaded region is 76π.

Therefore, the ratio between the shaded and unshaded region is $24\pi/76\pi = 6/19$.

The answer is (B).

Solution 2:

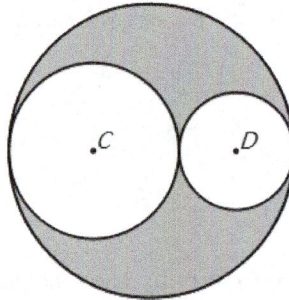

The original shaded region is half of what is shown in the diagram above whose area is $100\pi - 16\pi - 36\pi = 48\pi$. Thus the area of the origin region is 24π.

Then, the unshaded area is $100\pi - 24\pi = 76\pi$, so the ratio between the shaded and unshaded region is $(24\pi)/(76\pi) = 6/19$.

The answer is (B).

Problem 104

In triangle ABC, $\angle B = 90°$. $BC = 6$, $AC = 8$. Find the area of the regions outside the circle but inside the triangle.

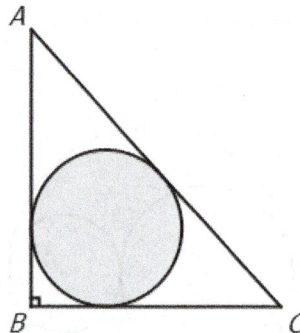

Tips: *Find the diameter of the inscribed circle.*

Solution:

176

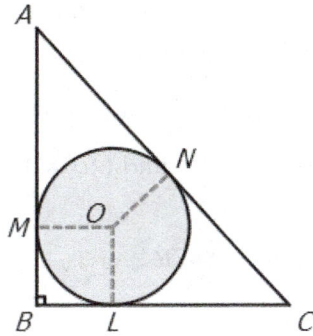

In the right $\triangle ABC$, $AC = 10$ and the diameter of the inscribed circle is equal to $AB + BC - AC = 6 + 8 - 10 = 4$.

The area of $\triangle ABC$ is 24 and the area of the shaded region is 4π, thus the area of the region outside the circle is $24 - 4\pi$.

Problem 105

$\triangle ABC$ is an equilateral triangle and O is the center of its inscribed circle. If the area of the circle is π, what is the area of $\triangle ABC$?

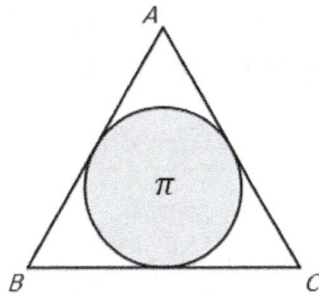

> **Tips:** *1. The incenter of an equilateral triangle is also the centroid of the triangle. 2. Use the properties of an equilateral triangle to find lengths.*

Solution 1:

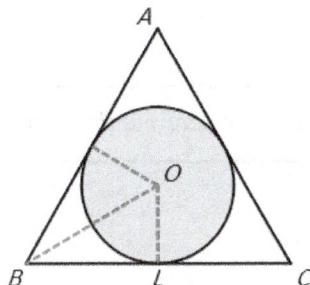

Let L be the midpoint of side BC and connect OL and OB to form a right triangle with $\angle OBL = 30°$.

The area of the inscribed circle is π, which leads to its radius $OL = 1$.

Because in the right $\triangle OBL$, $\angle OBL = 30°$ and $OL = 1$, $BL = \sqrt{3}$. Therefore, the edge-length of the equilaterial triangle is $BC = 2\sqrt{3}$.

The area of the equilateral triangle is $\frac{\sqrt{3}}{4}BC^2 = 3\sqrt{3}$.

Solution 2:

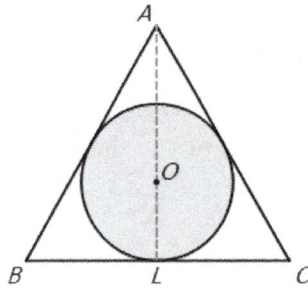

The area of the inscribed circle is π, which leads to its radius $OL = 1$.

Because the incenter O of an equilaterial triangle is also the circumcenter, $OL = \frac{1}{3}AL$ and $AL = 3$.

In the right triangle ABL, $AL/AB = \sqrt{3}/2$, thus $AB = 2\sqrt{3}$. Therefore, the area of the equilateral triangle is $\frac{\sqrt{3}}{4}AB^2 = 3\sqrt{3}$.

Problem 106

In an isosceles triangle ABC, $AB = AC = 7$ and $BC = 10$. What is the area of its inscribed circle?

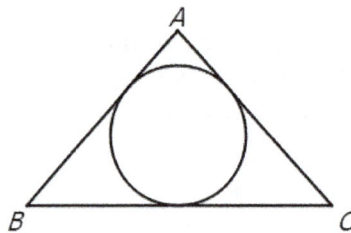

> **Tips:** *Find the radius of the inscribed circle through properties of similar right triangles.*

Solution 1:

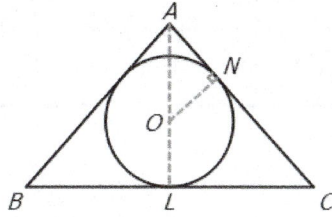

Let AL be the altitude to side BC and connect O, A and N, where O is the center of the circle and N is the tangent point. Because $BC = 10$ and $AC = 7$, $AL = 2\sqrt{6}$ and $LC = 5$.

Let radius of the circle $r = ON = OL$. Because $\triangle OAN \sim \triangle LAC$, $ON/LC = OA/AC$. Thus $r/5 = (2\sqrt{6} - r)/7$.

Solving the equation gives $r = 5/\sqrt{6}$ and the area of the circle as $\frac{25}{6}\pi$.

Solution 2:

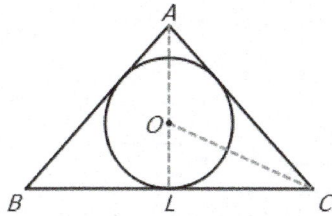

Let AL be the altitude to side BC. Because $BC = 10$ and $AC = 7$, we have $AL = 2\sqrt{6}$ and $LC = 5$.

Connect OC, which is the angular bisector of $\angle ACB$.

By the **Angular Bisector Theorem**, $OL/OA = LC/AC$. Thus $OL = AL*LC/(LC+AC) = 2\sqrt{6} * 5/(5+7) = 5/\sqrt{6}$.

Therefore, the area of the inscribed circle is $\frac{25}{6}\pi$.

Problem 107

An isosceles triangle ABC with $AB = AC = 7$ and $BC = 10$ is inscribed in a circle. What is the area of circle?

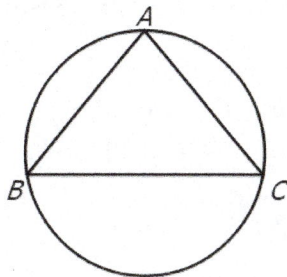

> **Tips:** *Find the radius of the circumscribed circle using the Pythagorean Theorem, or the law of sines, or the formula for the radius.*

Solution 1:

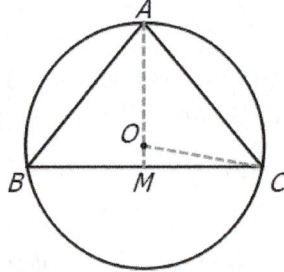

Let AM be the altitude to side BC and connect center of the circle, O, to vertex C. Because $BC = 10$ and $AC = 7$, $AM = 2\sqrt{6}$ and $MC = 5$.

Let radius of the circle $r = OA = OC$. In $Rt\triangle OMC$, by the **Pythagorean Theorem**, $OM = \sqrt{r^2 - 25}$.

$AM = OA + OM$ leads to $2\sqrt{6} = r + \sqrt{r^2 - 25}$.

Solving the equation gives $r = 49/(4\sqrt{6})$ and the area of the circle is $\frac{2401}{144}\pi$.

Solution 2:

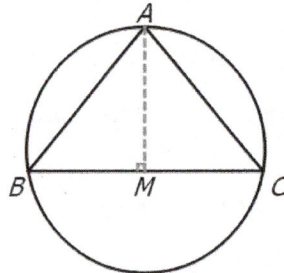

Let AM be the altitude to side BC. Because $BC = 10$ and $AC = 7$, $AM = 2\sqrt{6}$ and $MC = 5 \sin\angle B = AM/AB = 2\sqrt{6}/7$.

By the law of the sines, $2r = AC/sin(\angle B) = 49/(2\sqrt{6})$.

Therefore, $r = 49/(4\sqrt{6})$ and the area of the circle is $\frac{2401}{144}\pi$.

Solution 3:

The radius formula for circumscribed circles states that $r = \frac{abc}{4S_{\triangle ABC}}$. The area of $\triangle ABC$ is $10 * \sqrt{7^2 - 5^2}/2 = 10\sqrt{6}$, thus $r = \frac{7*7*10}{4*10\sqrt{6}} = 49/(4\sqrt{6})$.

Therefore, the area of the circle is $\frac{2401}{144}\pi$.

Problem 108

An isosceles triangle ABC with $AB = AC$ and BC is the diameter of the triangle's circumcircle. E and G are on the extension of AB and AC, respectively, such that $BC // EG$. Find the ratio of the area of triangle ABC and trapazoid $BCGE$?

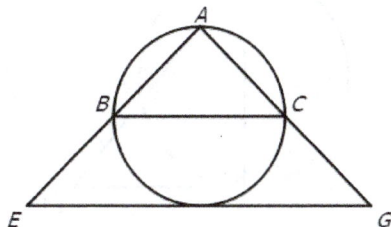

> **Tips:** *Construct the altitude of triangle ABC and AEG.*

Solution:

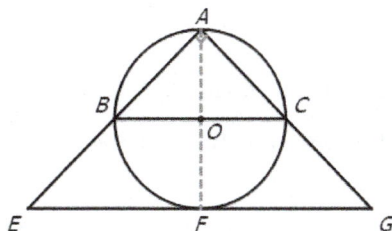

Because $AB = AC$ and BC is a diameter of the circle, $OA \perp BC$ and OA is the radius.

AF is the diameter of the circle, thus $OA/AF = 1/2$.

Because $BC // EG$, $\triangle EBC \sim \triangle EFD$ and $S_{\triangle ABC}/S_{\triangle AEG} = (OA/AF)^2 = 1/4$. Therefore, the ratio of the area of triangle ABC to the area of trapezoid $BCGE$ is $1/3$.

Problem 109

In the $Rt\triangle ABC$, $\angle B = 90°$, $AB = 8$ and $BC = 6$. Side AB is the diameter of the circle O and side AC meets the circle O at E. Find CE.

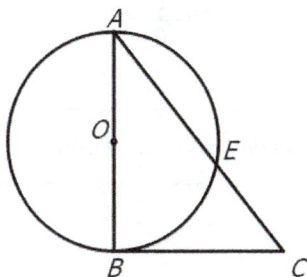

181

Tips: *Connect BE to form a right triangle and find similar triangle pairs.*

Solution 1:

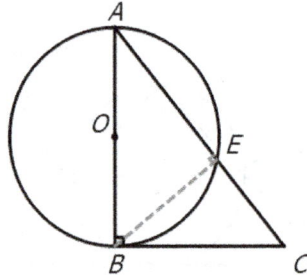

In the $Rt\triangle ABC$, $AB = 8$ and $BC = 6$, thus $AC = 10$.

Connect BE. Because AB is the diameter of the circle, $\angle AEB = 90°$. BE is the altitude to the hypotenuse of a right triangle, thus $AB^2 = AE * AC$ and $AE = AB^2/AC = 64/10$.

Therefore, $CE = AC - AE = 10 - 64/10 = 3.6$.

Solution 2:

In the $Rt\triangle ABC$, $AB = 8$ and $BC = 6$, thus $AC = 10$.

By the **Power of a Point Theorem**, $CB^2 = CE * AC$, thus $CE = CB^2/AC = 36/10 = 3.6$.

Problem 110

In the figure, AB and CD are diameters of the circle with center O, $AB \perp CD$, and chord DF intersects AB at E. If $DE = 6$ and $EF = 2$, then the area of the circle is

(A) 23π (B) $47\pi/2$ (C) 24π (D) $49\pi/2$ (E) 25π

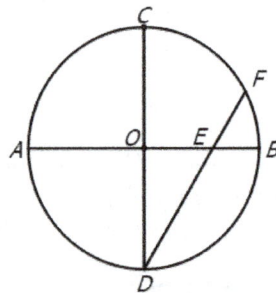

Tips: *Connect CF to form a right triangle similar to triangle ODE.*

Solution 1:

182

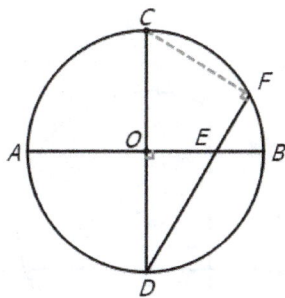

Connect CF and let $r = OD = OC$.

Because $\angle CDF = \angle EDO = 90°$, $\triangle CDF \sim \triangle EDO$. Then we know that $OD/DF = DE/CD$, which gives us $r/8 = 6/(2r)$. Solving the equation gives $r^2 = 24$ and the area of the circle as 24π.

The answer is (C).

Solution 2:

Let r be the radius of the circle and $x = OE$.

By the **Power of a Point Theorem**, $BE * EA = DE * EF$. Thus $(r - x)(r + x) = 2 * 6$, which leads to $r^2 - x^2 = 12$.

In the right triangle ODE, by the **Pythagorean Theorem**, $x^2 + r^2 = 36$. Solving the equation set gives $r^2 = 24$, thus the area of the circle is 24π.

The answer is (C).

Problem 111

Chords AB and CD in the circle intersect at E and are perpendicular to each other. If segments AE, EB, and ED have measures 2, 6, and 3 respectively, then the length of the diameter of the circle is

(A) $4\sqrt{5}$ (B) $\sqrt{65}$ (C) $2\sqrt{17}$ (D) $3\sqrt{7}$ (E) $6\sqrt{2}$

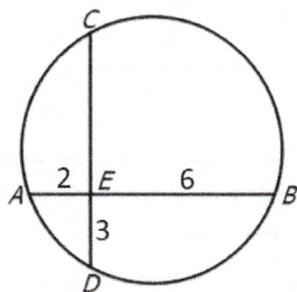

Tips: *Connect the center of the circle to the midpoints of the chords to form a rectangle.*

183

Solution:

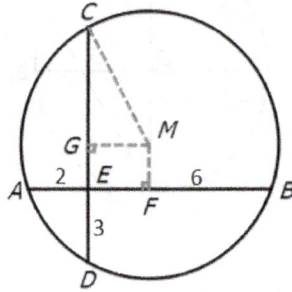

Connect center of the circle to midpoint of the two chords as shown in the diagram. MF is the perpendicular bisector of AB and MG is the perpendicular bisector of CD.

Thus $AF = FB = 4$, $GM = EF = AF - AE = 4 - 2 = 2$. By the **Power of a Point Theorem**, $CE * ED = AE * EB$, thus $CE = 2 * 6/3 = 4$ and $CG = (3+4)/2 = 7/2$.

In the $Rt\triangle CMG$, by the **Pythagorean Theorem**, $CM^2 = CG^2 + GM^2 = 49/4 + 4$. Therefore, the length of the diameter of the circle is $2CM = \sqrt{65}$.

The answer is (B).

Problem 112

Let $\triangle XOY$ be a right triangle with $\angle XOY = 90°$. Let M and N be the midpoints of legs OX and OY, respectively. Suppose that $XN = 19$ and $YM = 22$. What is XY ? (2002 AMC10B)

(A) 24   (B) 26   (C) 28   (D) 30   (E) 32

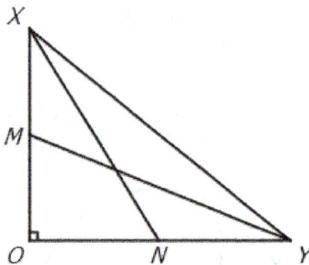

> **Tips:** *Apply the Pythagorean Theorem to all three right triangles in the problem.*

Solution 1:

184

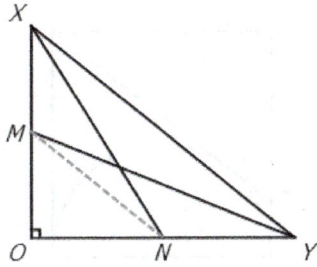

MN is a midline and $MN = XY/2$. By the **Pythagorean Theorem**, $XN^2 + YM^2 = OX^2 + ON^2 + OY^2 + OM^2 = (OX^2 + OY^2) + (ON^2 + OM^2) = MN^2 + XY^2 = (5/4)XY^2$.

Thus, $XY^2 = (4/5) * (19^2 + 22^2) = 26^2$ and $XY = 26$.

The answer is (B).

Solution 2:

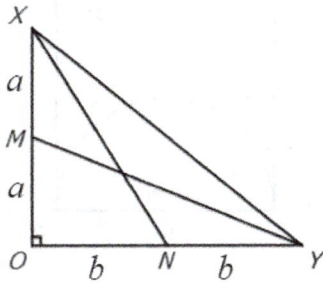

Let $2b = OY$ and $2a = OX$. Thus $ON = b$ and $OM = a$.

Applying the **Pythagorean Theorem** in triangles OXN and OYM gives $XN^2 = b^2 + 4a^2$, $YM^2 = a^2 + 4b^2$. Thus $XN^2 + YM^2 = 5(a^2 + b^2)$.

Therefore, in the right triangle OXY, $XY^2 = 4(a^2 + b^2) = (4/5) * (XN^2 + YM^2) = (4/5) * (19^2 + 22^2) = 26^2$ and $XY = 26$.

The answer is (B).

Problem 113

Square $ABCD$ has side length 2. A semicircle with diameter AB is constructed inside the square, and the tangent to the semicircle from C intersects side AD at E. What is the length of CE?

(A) $(2 + \sqrt{5})/2$ (B) $\sqrt{5}$ (C) $\sqrt{6}$ (D) $5/2$ (E) $5 - \sqrt{5}$

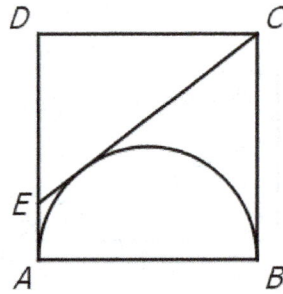

Tips: *Connect center of the circle to tangent point and apply the Pythagorean Theorem to the right triangles formed.*

Solution 1:

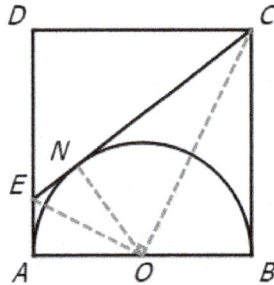

Because CE, AE, and CB are tangent to the semicircle, $\angle AOE = \angle NOE$, $\angle BOC = \angle NOC$, $CN = BC = 2$ and $ON \perp CE$. Thus $\angle COE = 90°$.

In the right triangle COE, $ON \perp CE$. Thus $ON^2 = CN * NE$ and $NE = ON^2/CN = 1/2$. Therefore, $CE = CN + NE = 2 + 1/2 = 5/2$.

The answer is (D).

Solution 2:

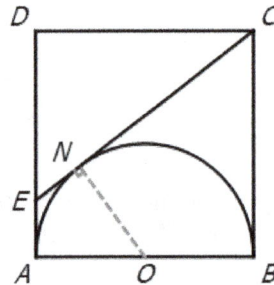

Because CE, AE and CB are tangent to the semicircle, $CN = BC = 2$ and $NE = EA$.

Let $x = NE = EA$. In the right triangle CDE, $DE = 2 - x$, $CE = 2 + x$ and $CD = 2$. By the **Pythagorean Theorem**, $CE^2 = DE^2 + CD^2$, so $(2 + x)^2 = (2 - x)^2 + 4$.

Solving the equation gives $x = 1/2$. Therefore, $CE = 1/2 + 2 = 5/2$.

Answer is (D).

Problem 114

In the $\triangle ABC$, $AB = 7$, $AC = 11$ and $BD = 14$, respectively. Find the length of the altitude of the triangle drawn to side BC.

(A) $14\sqrt{10}/7$ (B) $12\sqrt{5}/7$ (C) $31/7$ (D) $12\sqrt{10}/7$ (E) $12\sqrt{10}/49$

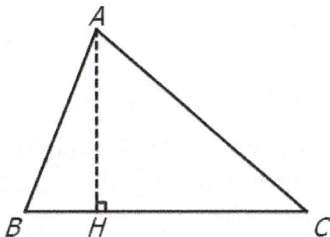

> **Tips:** *1. Use Heron's Formula to calculate triangle area, or 2. Use trigonometry to calculate $\cos(B)$ and then $\sin(B)$.*

Solution 1:

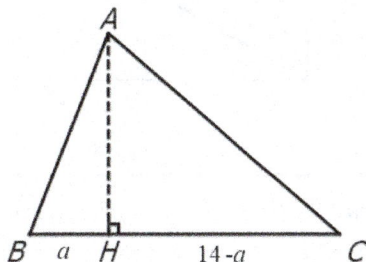

Construct the altitude AH on BC. Applying the **Pythagorean Theorem** to right triangle ABH and ACH gives $AH^2 = AB^2 - a^2$ and $AH^2 = AC^2 - (14 - a)^2$. Thus $7^2 - a^2 = 11^2 - (14 - a)^2$.

Solving the equation gives $a = 31/7$. Therefore, $AH = 12\sqrt{10}/7$.

The answer is (D).

Solution 2:

Calculate the area through **Heron's Formula**. We obtain $s = (7 + 11 + 14)/2 = 16$, thus the triangle area is $S_{\triangle ABC} = \sqrt{s(s-7)(s-11)(s-14)} = 12\sqrt{10}$.

The triangle area is also equal to $\frac{1}{2}AH * BC$, hence $AH = 2S_{\triangle ABC}/BC = 12\sqrt{10}/7$.

The answer is (D).

Solution 3:

187

By the law of cosines,

$$\cos \angle B = \frac{AB^2 + BC^2 - AC^2}{2AB * BC} = \frac{31}{49}$$

Thus $\sin \angle B = \frac{12\sqrt{10}}{49}$. Therefore,

$$AH = AB \sin \angle B = 7 * \frac{12\sqrt{10}}{49} = 12\sqrt{10}/7$$

The answer is (D).

Problem 115

In the right triangle ABC, E and D are the trisection points of the hypotenuse AB. If $CD = 7$ and $CE = 6$, what is the length of hypotenuse AB?

(A) $\sqrt{17}$ (B) $3\sqrt{17}$ (C) $3\sqrt{17}/2$ (D) 5   (E) 4

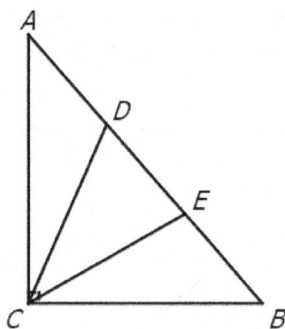

Tips: 1. Construct a rectangle to create congruent and similar triangles.

Solution 1:

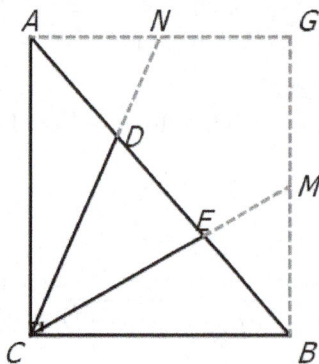

Construct a rectangle $ACBG$ based on the right triangle ACB as shown in the diagram. CD intersects AG at point N and CE intersects GB at point M.

188

Because $AG//CG$ and $BD = 2AD$, $BC = 2AN$ and $CD = 2ND$. Similarly, because $AC//GB$ and $AE = 2EB$, $AC = 2MB$ and $CE = 2EM$.

By the **Pythagorean Theorem**, $CN^2 = AN^2 + AC^2$ and $CM^2 = MB^2 + BC^2$. So $CN^2 + CM^2 = AN^2 + BC^2 + AC^2 + MB^2 = (5/4)(AC^2 + BC^2) = (5/4)AB^2$.

With $CN = (3/2)CD = 21/2$ and $CM = (3/2)CE = 9$, $AB^2 = (4/5)(CN^2 + CM^2) = (4/5)(441/4 + 81) = 153$. Thus $AB = 3\sqrt{17}$.

The answer is (B).

Solution 2:

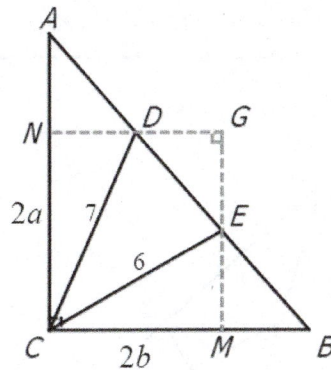

Construct a rectangle $NGMC$, such that NG passes through point D and GM passes through point E. Because $NG//BC$, $MG//AC$ and $AD = DE = EB$, $ND = DG = CM/2$ and $EM = EG = NC/2$.

Let $2b = CN$ and $2a = CM$. Thus $AN = a$, $BM = b$, $ND = DG = b$ and $EM = EG = a$.

In the rectangle, by the **Pythagorean Theorem**, $CD^2 = DN^2 + CN^2$ and $CE^2 = CM^2 + EM^2$. Hence $CD^2 + CE^2 = DN^2 + CM^2 + NC^2 + EM^2 = 5(a^2 + b^2)$. Therefore, $a^2 + b^2 = 17 = DE^2$ and $AB = 3DE = 3\sqrt{17}$.

Answer is (B).

Problem 116

In $\triangle ABC$, $BC = 30$ and $AC = 20$. AD and BE are two medians on BC and AC, respectively, with $AD \perp BE$. Find the length of AB.

(A) $2\sqrt{65}$ (B) 15   (C) 65   (D) 16   (E) $10\sqrt{5}$

189

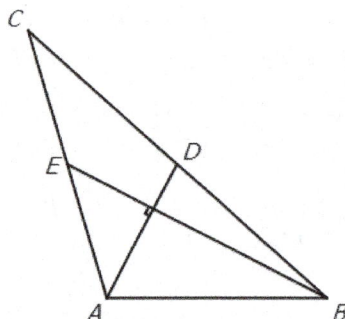

Tips: *Connect DE and use the Pythagorean Theorem in the four right triangles.*

Solution:

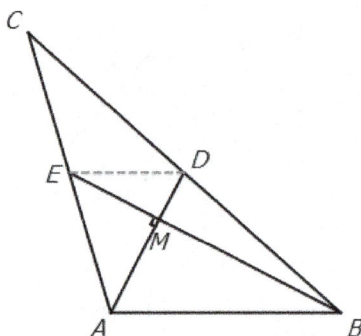

Connect DE. The midline $DE = AB/2$ and $AE = 10$ and $BD = 15$.

By the **Pythagorean Theorem**, $AE^2 = EM^2 + AM^2$, $BD^2 = DM^2 + BM^2$. Thus, $AE^2 + BD^2 = EM^2 + AM^2 + DM^2 + BM^2 = (EM^2 + DM^2) + (AM^2 + BM^2) = DE^2 + AB^2 = (5/4)AB^2$.

Therefore, $AB^2 = (4/5)(AE^2 + BD^2) = 260$ and $AB = 2\sqrt{65}$.

The answer is (A).

Problem 117

Points A, B, C, and D lie on a line, in that order, with $AB = CD$ and $BC = 12$. Point E is not on the line, and $BE = CE = 10$. The perimeter of $\triangle AED$ is twice the perimeter of $\triangle BEC$. Find AB. (2002 AMC10A)

(A) 15/2   (B) 8   (C) 17/2   (D) 9   (E) 19/2

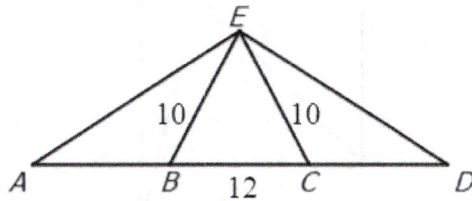

Tips: *Construct right triangles through the altitude of triangle EBC on side BC and use the Pythagorean Theorem.*

Solution:

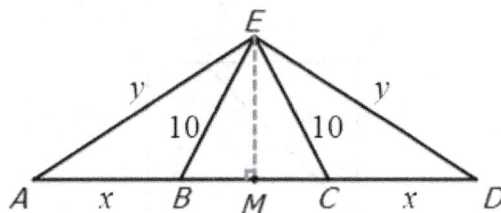

Let M be the midpoint of BC and connect EM. Then, $BM = 6$ and $EM = 8$. Let $x = AB = CD$ and $y = AE = DE$.

By the **Pythagorean Theorem**, in the right triangle AME, $y^2 = (x+6)^2 + 8^2$. The perimeter constraint gives $2y + 2x + 12 = 2 * (10 + 10 + 12)$. We are left with the equation set $y^2 = (x+6)^2 + 8^2$ and $(x+6) + y = 32$.

Solving the equation set gives uss $x + 6 = 15$ and thus $x = 9$.

The answer is (D).

Problem 118

Points E and F are located on square $ABCD$ so that $\triangle BEF$ is equilateral. What is the ratio of the area of $\triangle DEF$ to that of $\triangle ABE$? (2004 AMC10A)

(A) 4/3   (B) 3/2   (C) $\sqrt{3}$ (D) 2   (E) $1 + \sqrt{3}$

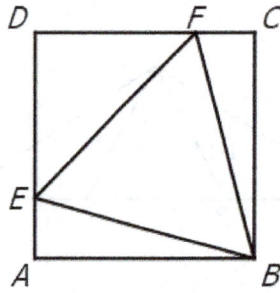

Tips: *Connect BD and use the Pythagorean Theorem to establish a constraint about segment AE.*

Solution:

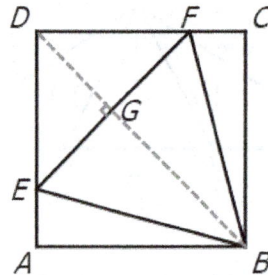

Connect BD and label the intersection with EF as point G. Let the side-length of the square be 1 and length of AE be x. Hence $DE = 1 - x$ and $EG = (\sqrt{2}/2)DE = (\sqrt{2}/2)(1 - x)$.

In the equilateral triangle, $BE = 2EG = \sqrt{2}(1 - x)$. By the **Pythagorean Theorem** in triangle $\triangle ABE$, $BE^2 = AE^2 + AB^2$. So $2(1 - x)^2 = x^2 + 1$. Solving the quadratic equation gives $x = 2 - \sqrt{3}$.

The area of $\triangle DEF$ is $(1 - x)^2/2 = 2 - \sqrt{3}$ and the area of $\triangle ABE$ is $x/2 = (2 - \sqrt{3})/2$. Therefore, the ratio of the area of $\triangle DEF$ to that of $\triangle ABE$ is 2.

The answer is (D).

Problem 119

In $\triangle ABC$, we have $AB = AC = 61$ and $BC = 22$. D is a point on the extension of BC such that $AD = 100$. What is CD?

(A) 22   (B) 42   (C) 52   (D) 69   (E) 64

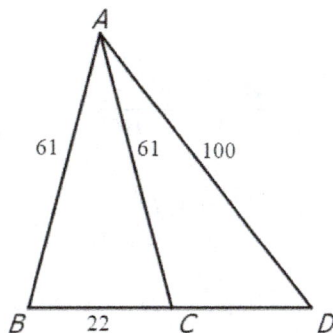

Tips: *Construct the altitude on side BC to form a right triangle involving side AD.*

Solution 1:

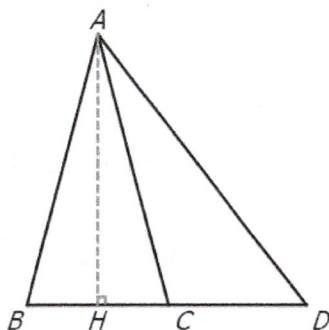

Let A be a point on BD such that $AH \perp BC$. AH is the altitude of the isosceles triangle, so $HC = BC/2 = 11$.

By the **Pythagorean Theorem**, $AH = \sqrt{AH^2 - HC^2} = \sqrt{61^2 - 11^2} = 60$ and $HD = \sqrt{AD^2 - AH^2} = \sqrt{100^2 - 60^2} = 80$.

Therefore, $CD = HD - HC = 80 - 11 = 69$.

The answer is (D).

Solution 2:

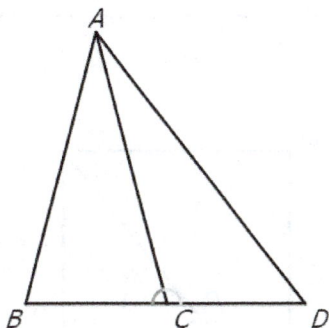

Let $x = CD$. Next, use the law of cosines to calculation $\cos \angle ACB$ and $\cos \angle ACD$.

193

The law of cosines gives:

$$\cos \angle ACB = \frac{AB^2 + BC^2 - AC^2}{2AB * BC} = \frac{22^2}{2 * 22 * 61} = \frac{11}{61}$$

and

$$cos(\angle ACD) = \frac{AC^2 + CD^2 - AD^2}{2AC * CD} = \frac{x^2 + 61^2 - 100^2}{2x * 61}$$

Because $\angle ACB + \angle ACD = 180°$, $cos(\angle ACB) + cos(\angle ACD) = 0$. Thus

$$\frac{x^2 + 61^2 - 100^2}{2x * 61} + \frac{11}{61} = 0$$

Solving the equation gives $x = 69$.

The answer is (D).

Problem 120

Point P is inside square $ABCD$ such that $PA = 5$, $PB = 8$, and $PC = 13$. Find the area of the square.

(A) 153   (B) 126   (C) 128   (D) 130   (E) 132

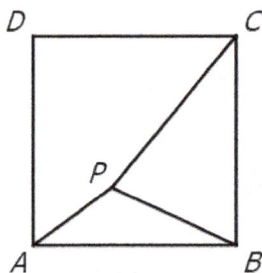

Tips: *Construct right triangles through the altitudes on side AB and BC to establish equations.*

Solution 1:

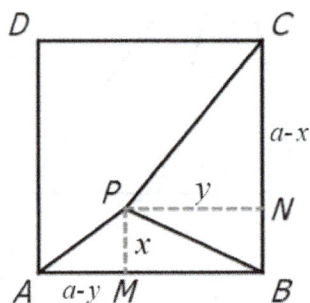

194

Construct rectangle $PMBN$ as shown in the diagram. Let $x = PM = NB$, $y = PN = MB$, and $a = AB = BC$. Thus $AM = a - y$ and $CN = a - x$.

By the **Pythagorean Theorem**, in the triangle PAM, PMB and PCN, we have: $PA^2 = 25 = x^2 + (a-y)^2$, $PB^2 = 64 = x^2 + y^2$, $PC^2 = 169 = y^2 + (a-x)^2$

Solving the quadratic equation set gives $a^2 = 153$ or 41 (ignore).

The answer is (A).

Solution 2:

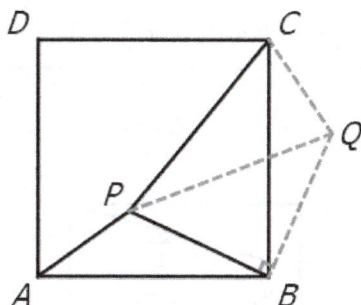

We can also use the law of cosines to find BC^2, which is the area of the square.

Rotate triangle PAB $90°$ clockwise around vertext B to form QBC. Thus $QB = PB = 8$, $CQ = PA = 5$ and $\angle PBQ = 90°$.

In the isosceles right triangle PBQ, $PQ = 8\sqrt{2}$. By the law of cosines, in triangle PCQ, we have

$$\cos \angle CPQ = \frac{PC^2 + PQ^2 - QC^2}{2PC * PQ} = \frac{13^2 + 128 - 25}{2 * 13 * 8\sqrt{2}} = \frac{17}{13\sqrt{2}}$$

Therefore,

$$\sin \angle CPQ = \frac{7}{13\sqrt{2}}$$

$$\cos \angle BPC = \cos \angle CPQ + 45° = \frac{5}{13}$$

By the law of cosines, $BC^2 = PC^2 + PB^2 - 2PC*PB*\cos \angle BPC = 13^2 + 8^2 - 2*13*8*\frac{5}{13} = 153$.

The answer is (D).

Problem 121

If point Q lies on side AD of square $ABCD$ such that $QC = \sqrt{10}$ units and $QB = \sqrt{13}$ units, what is the area of square $ABCD$? (2015 Mathcounts)

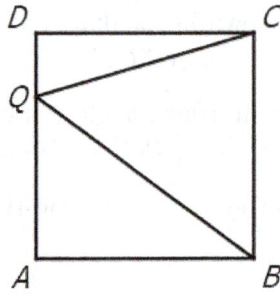

Tips: *Use the Pythagorean Theorem in right triangle QCD and QAB.*

Solution:

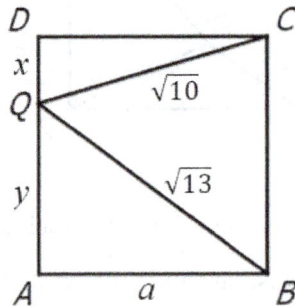

Let $x = DQ$, $y = AQ$ and $a = AB$. By the **Pythagorean Theorem**, in triangle QCD and QAB, $10 = x^2 + a^2$ and $13 = y^2 + a^2$. In the square we also have $x + y = a$. The equation set is: $x^2 + a^2 = 10$, $y^2 + a^2 = 13$, and $x + y = a$

Solving the equation set gives $x^2 = 1$, $y^2 = 4$, and $a^2 = 9$, thus the area of the square is 9.

Problem 122

A square $ABCD$ has line segments drawn from vertex B to the midpoints N and M of sides AD and DC respectively. Find the ratio of the perimeter of quadrilateral $BMDN$ to the perimeter of square $ABCD$. (Mathcounts)

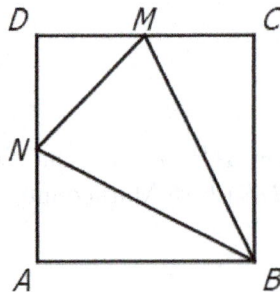

196

Tips: *Use the Pythagorean Theorem to calculate the length of BM and BN.*

Solution:

Let the side-length of AB be 2. Then the perimeter of the square $ABCD$ is 8.

By the **Pythagorean Theorem**, in the right triangle ABN and BCM, $BM = BN = \sqrt{5}$, thus the perimeter of $BMDN$ is $2 + 2\sqrt{5}$. Therefore, the ratio of the perimeter of $BMDN$ to the perimeter of square $ABCD$ is $(1 + \sqrt{5})/4$.

Problem 123

A quadrilateral $ABCD$ with $AB = 4$, $BC = 6$, $\angle DAB = 120°$, $\angle DCB = 60°$ and $\angle B = 90°$. Find AD.

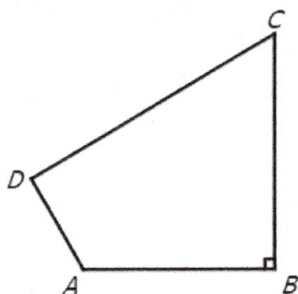

Tips: *Construct right triangles and use the Pythagorean Theorem.*

Solution 1:

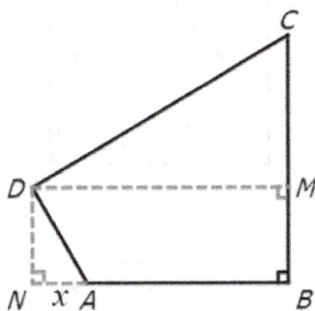

Construct a rectangle $BMDN$ as show in the diagram and let $x = AN$.

In the right $\triangle ADN$, $\angle DAN = 180° - \angle DAB = 180° - 120° = 60°$. So $DN = \sqrt{3}AN = \sqrt{3}x$. Similarly, in the right $\triangle CDM$, $\angle CMD = 60°$. Thus $DM = \sqrt{3}CM$.

Because $DM = BN = x + 4$ and $CM = BC - BM = 6 - \sqrt{3}x$, we obtain $x + 4 = \sqrt{3}(6 - \sqrt{3}x)$.

Solving the equation gives $x = (3/2) * \sqrt{3} - 1$ and $AD = 2x = 3\sqrt{3} - 2$.

Solution 2:

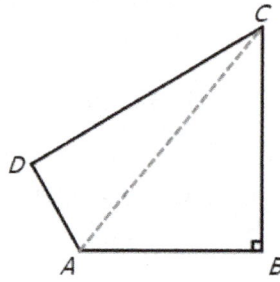

In this solution, we use trigonometry to calculate the length of AD. Because $\angle DAB + \angle DCB = 180°$, $\angle ADC = 90°$.

In the $Rt\triangle ABC$, $AC = \sqrt{AB^2 + BC^2} = \sqrt{4^2 + 6^2} = 2\sqrt{13}$ and $\sin\angle ACB = 2/\sqrt{13}$ and $\cos\angle ACB = 3/\sqrt{13}$.

Using trigonometric identities, we have $\sin\angle ACD = \sin(60° - \angle ACB) = \sin 60° \cos\angle ACB - \cos 60° \sin\angle ACB = (3\sqrt{3} - 2)/(2\sqrt{13})$.

Therefore, $AD = AC * \sin\angle ACD = 2\sqrt{13} * (3\sqrt{3} - 2)/(2\sqrt{13}) = 3\sqrt{3} - 2$.

Problem 124

The area of the largest equilateral triangle that can be inscribed in a square of side length 1 unit can be expressed in the form $a\sqrt{b} - c$ $units^2$, where a, b and c are integers. What is the value of $a + b + c$? (2014 Mathcounts National Sprint)

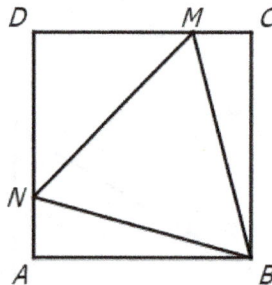

> **Tips:** *Express the edge-length of the equilateral triangle through Pythagorean Theorem.*

Solution 1:

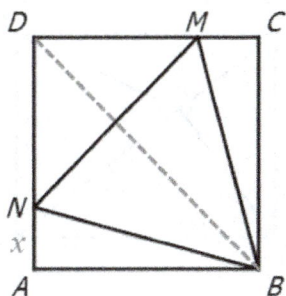

Let $x = NA$. In the isosceles triangle DMN, $DN = 1 - x$ and $MN = \sqrt{2}(1 - x)$.

In the right $\triangle ABN$, $NB = \sqrt{1 + x^2}$. Because $MN = NB$, $\sqrt{2}(1 - x) = \sqrt{1 + x^2}$.

Solving the equation gives us $x = 2 - \sqrt{3}$ and the area of equilateral triangle as $\sqrt{3}/4 * (1 + x^2) = (\sqrt{3}/4) * (1 + 7 - 4\sqrt{3}) = 2\sqrt{3} - 3$. Therefore, $a + b + c = 8$.

Solution 2:

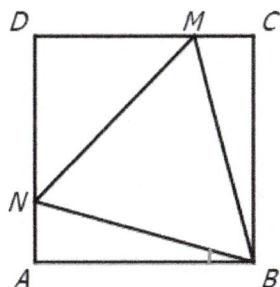

We know $\angle ABN = 15°$ and we want to find the value of $\cos 15°$. Using trigonometric identities, we have

$$\cos 15° = \cos(45° - 30°) = \cos 45° \cos 30° + \sin 45° \sin 30° = \frac{\sqrt{6} + \sqrt{2}}{4}$$

Thus

$$BN = \frac{AB}{\cos 15°} = \frac{1}{(\frac{\sqrt{6}+\sqrt{2}}{4}))} = \sqrt{6} - \sqrt{2}$$

Therefore, the area of equilateral triangle is $(\sqrt{3}/4) * (\sqrt{6} - \sqrt{2})^2 = 2\sqrt{3} - 3$, and $a + b + c = 8$.

Problem 125

In rectangle $ABCD$, the diagonal $BD = 25$, and $BC/CD = 3/4$. H is a point on line BD such that $AH \perp BD$. Connect CH. What is the area of $\triangle BCG$?

(A) 96   (B) 110   (C) 115   (D) 120   (E) 130

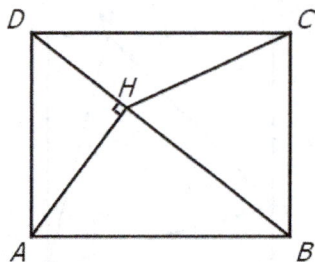

Tips: *Calculate length of DH through the properties of similar triangles.*

Solution:

Because in the right triangle BCD, $BC/CD = 3/4$ and $BD = 25$, $BC = 15$ and $CD = 20$.

$\triangle BCD \sim \triangle DHA$ leads to $DH = (3/5) * AD = (3/5) * BC = 9$.

Therefore, $BH/BD = (25 - 9)/25 = 16/25$ and the area of $\triangle BCH = (BH/BD) * S_{\triangle ABC} = (16/25) * 15 * (20/2) = 96$.

The answer is (A).

Problem 126

In $\triangle ABC$, side AC and the perpendicular bisector of BC meet in point D, and BD bisects $\triangle ABC$. If $AD = 9$ and $DC = 7$, what is the area of $\triangle ABD$? (2002 AMC12A)

(A) 14   (B) 21   (C) 28   (D) $14\sqrt{5}$ (E) $28\sqrt{5}$

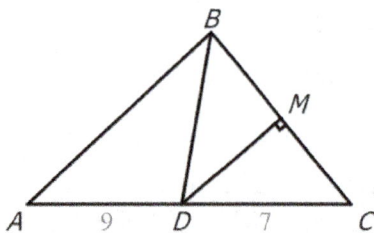

Tips: *Establish equations based on the angular bisector length formula.*

Solution 1:

Because DM is the perpendicular bisector of BC, $BD = CD = 7$. Since BD is the angular bisector of $\angle ABC$, $AB/BC = AD/CD = 9/7$. Let $AB = 9x$ and $BC = 7x$.

The length of angular bisector formula gives $BD^2 = AB * BC - AD * CD = 63x^2 - 63 = 7^2 = 49$.

Solving the equation yields $x = 4/3$ and $AB = 9x = 12$. In the $\triangle ABD$, $AB = 12$, $AD = 9$ and $BD = 7$. Using **Heron's Formula**, the area of $\triangle ABD$ is $\sqrt{14 * 2 * 5 * 7} = 14\sqrt{5}$.

200

The answer is (D).

Solution 2:

Similar to in the solution 1, we have $AB = 9$ and $BC = 28/3$.

Applying the **Pythagorean Theorem** in the right equation BCD yields $DM = \sqrt{7^2 - (14/3)^2} = (7/3) * \sqrt{5}$. Therefore, the area of $\triangle AEC$, $S_{\triangle BCD} = BC * DM/2 = (98/9) * \sqrt{5}$.

We also have $S_{\triangle ABD}/S_{\triangle BCD} = AB/BC = 9/7$, hence the area of $\triangle ABD$ is $(9/7) * (98/9) * \sqrt{5} = 14\sqrt{5}$.

The answer is (D).

Problem 127

In $\triangle ABC$, $AB = 6$ and $AC = 3$. Point D is on segment BC so that $BD : DC = 2 : 1$. If $AD = 2$, find the length of BC.

(A) $\sqrt{7}$ (B) $\sqrt{11}$ (C) $3\sqrt{7}$ (D) $3\sqrt{11}$ (E) $3\sqrt{18}$

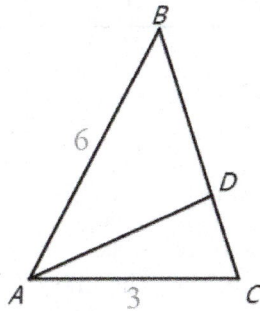

> **Tips:** *Use the converse of the angular bisector theorem and the angular bisector length theorem.*

Solution:

Because $AB/AC = 2 : 1 = BD/DC$, AD is the angular bisector of $\angle A$.

Let $BD = 2x$ and $DC = x$.

Applying the **Angular Bisector Length Theorem** yields $3 * 6 - x * 2x = 2^2$. Solving the equation set gives $x = \sqrt{7}$, thus BC is $3\sqrt{7}$.

The answer is (C).

Problem 128

The perimeter of the $\triangle ABC$ is 24 and M is the midpoint of AC and $MB = MA = 5$. What is the area of $\triangle ABC$?

(A) 12   (B) 16   (C) 24   (D) 30

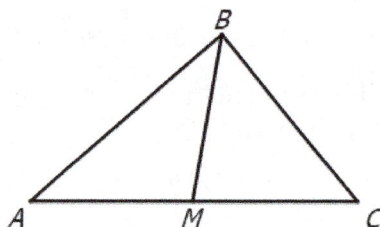

Tips: *Use converse of the Pythagorean Theorem.*

Solution:

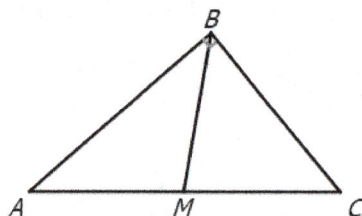

Because $MA = MC = MB$, $\triangle ABC$ is a right triangle and $\angle ABC = 90°$.

Let $x = AB$ and $y = BC$. The perimeter of the triangle is $10 + x + y = 24$.

Applying the **Pythagorean Theorem** yields $x^2 + y^2 = 100$. Solving the equation set gives $x = 6$ and $y = 8$ or $x = 8$ and $y = 6$, thus the area of is 24.

The answer is (C).

Problem 129

In $\triangle ABC$, M is the midpoint of AC. ME and MF are the angle bisector of $\angle BMC$ and $\angle BMA$, respectively. Find $\angle BFE$ if $\angle BAC = 32°$.

(A) 32°   (B) 36°   (C) 38°   (D) 48°   (E) 58°

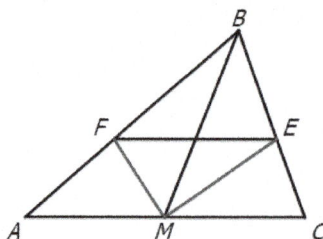

Tips: *Use angular bisector properties for two angular bisectors.*

Solution:

Because MF and ME are angular bisector of $\angle BMC$ and $\angle BMA$, $BM/MC = BE/EC$ and $BM/MA = BF/FA$.

Knowing $MC = MA$, we have $BE/EC = BF/FA$. Therefore, $EF//AC$ and $\angle BFE = \angle BAC = 32°$.

The answer is (A).

Problem 130

In $\triangle ABC$, the ratio $AC : CB = 3 : 4$. The bisector of the exterior angle at C intersects BA extended at P (A is between P and B). The ratio $PA : AB$ is:

(A) $3/3$ (B) $3/4$ (C) $4/3$ (D) $3/1$ (E) $7/1$

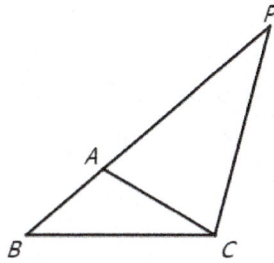

Tips: *Angular bisector leads to using the angle to calculate area of two triangles ACP and ABP.*

Solution 1:

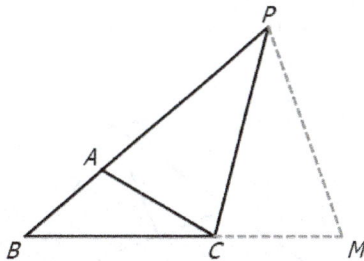

Extend BC to M such that $CM = AC$. Because $CM = AC$ and $\angle ACP = \angle MCP$, $\triangle ACP \cong \triangle MCP$. Thus $PM = PA$, $CM = AC$ and $\angle APC = \angle MPC$.

Because $\angle APC = \angle MPC$, PC is the angular bisector of $\angle BPM$. Then $PM/PB = CM/BC = AC/BC = 3/4$, thus $PA/PB = 3/4$ and $PA/AB = 3/1$.

The answer is (D).

Solution 2:

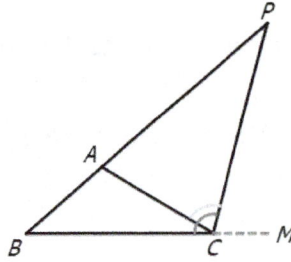

Because PC is the angular bisector of $\angle ACM$, $\angle BCP + \angle ACP = \angle BCP + \angle PCM = 180°$. Therefore, $\sin \angle BCP = \sin \angle ACP$.

The ratio of area of $\triangle ACP$ and $\triangle BCP$ has

$$\frac{S_{\triangle ACP}}{S_{\triangle BCP}} = \frac{PA}{PB}$$

On the other hand, the area of triangle $\triangle ACP = AC * CP * \sin \angle ACP$ and $S_{\triangle BCP} = BC * CP * \sin \angle BCP$. Hence

$$\frac{S_{\triangle ACP}}{S_{\triangle BCP}} = \frac{AC}{BC} = \frac{PA}{PB} = 3/4$$

Therefore, $PA/AB = (3/4)/(1 - 3/4) = 3/1$.

The answer is (D).

Problem 131

Medians BD and CE of a $\triangle ABC$ are perpendicular, $BD = 8$, and $CE = 12$. Find the area of triangle ABC.

(A) 24   (B) 32   (C) 48   (D) 64   (E) 96

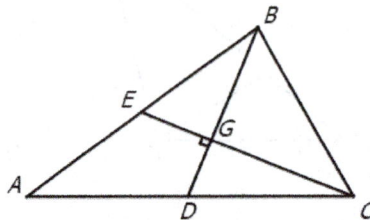

> **Tips:** *Use the properties of the intersection of the medians (centroid), i.e. 1/3 split.*

204

Solution 1:

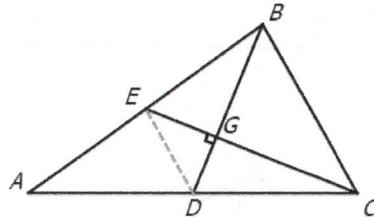

Connect DE, which is the midline of triangle ABC. Therefore, the area of triangle ADE is $1/4$ of the area of triangle ABC and the area of $BCDE$ is $3/4$ of the area of triangle ABC. Because in the quadrilateral $BCDE$, $BC \perp DE$, the area of $BCDE$ is $(1/2) * BD * CE = (1/2) * 8 * 12 = 48$. The area of triangle ABC is $48/(3/4) = 64$.

The answer is (D).

Solution 2:

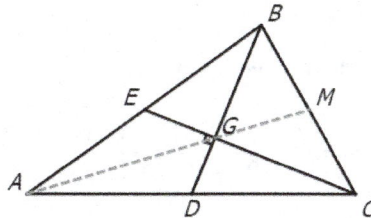

Let M be the midpoint of BC and connect AM.

Because G is the centroid of triangle ABC, $DG = (1/3)*BD = 8/3$ and $CG = (2/3)*CE = 8$ and the area of triangle CDG is $1/6$ of the area of the triangle ABC.

The area of triangle $CDG = (1/2) * DG * CG = (1/2) * (8/3) * 8 = 32/3$ and the area of triangle $ABC = 6 * (32/3) = 64$.

The answer is (D).

Problem 132

In the adjoining figure triangle ABC is such that $AB = 4$ and $AC = 8$. If M is the midpoint of BC and $AM = 3$, what is the length of BC?

(A) $2\sqrt{26}$ (B) $2\sqrt{31}$ (C) 9 (D) $4 + 2\sqrt{13}$ (E) not enough information to solve the problem

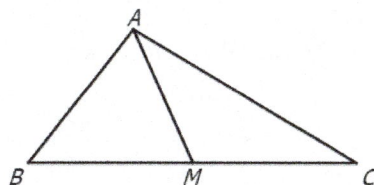

Tips: *Construct the altitude on side BC, use the law of cosines, or use the median length formula.*

Solution 1:

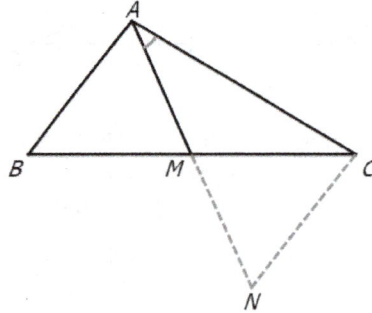

Extend AM to point N such that $MN = AM$ and connect NC. Because $BM = CM$ and $AM = MN$, $CN = AB = 4$ and $AN = 2AM = 6$.

Using the law of cosines in $\triangle ANC$, we have

$$\cos\angle CAM = \frac{AN^2 + AC^2 - CN^2}{2 \cdot AN \cdot AC} = \frac{36 + 64 - 16}{2 \times 6 \times 8} = \frac{7}{8}$$

In the triangle AMC, $MC^2 = AM^2 + AC^2 - 2AM*AC*\cos\angle CAM = 9 + 64 - 2*3*8*7/8 = 31$. Therefore, $BC = 2MC = 2\sqrt{31}$.

The answer is (B).

Solution 2:

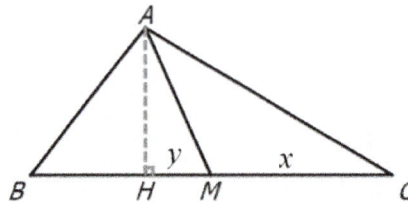

Let H be a point on BC such that $AH \perp BC$. Let $x = MC = MB$ and $y = HM$. Then $BH = x - y$ and $HC = x + y$.

By the **Pythagorean Theorem**, for the three right triangle ABH, AHM and ACH, we have $AH^2 = AB^2 - BH^2 = AM^2 - HM^2 = AC^2 - HC^2$, which gives $16 - (x - y)^2 = 9 - y^2$ and $9 - y^2 = 64 - (x + y)^2$.

Solving the equation set gives $x^2 = 31$. Therefore, $BC = 2x = 2\sqrt{31}$.

The answer is (B).

Solution 3:

Using the **Median Length Formula** yields $AM^2 + BM^2 = 1/2(AB^2 + AC^2)$, so $BM^2 = (4^2 + 8^2)/2 - 3^2 = 31$. Therefore, $BC = 2BM = 2\sqrt{31}$.

The answer is (B).

Problem 133

In the triangle ABC, $AB = 5$, $AC = 7$ and $BC = 8$. If AM is the median of edge BC, what is the length of AM?

(A) $\sqrt{26}$ (B) $\sqrt{31}$ (C) $\sqrt{21}$ (D) $2 + \sqrt{13}$ (E) $\sqrt{35}$

Tips: *Construct the altitude on side BC, use the law of cosines, or use the median length formula.*

Solution 1:

Using the law of cosines in $\triangle ABC$, we have

$$cos(\angle C) = \frac{BC^2 + AC^2 - AB^2}{2 \cdot BC \cdot AC} = \frac{64 + 49 - 25}{2 \times 7 \times 8} = \frac{11}{14}$$

Hence in the triangle AMC, $AM^2 = AC^2 + MC^2 - 2AC * MC * \cos \angle C = 49 + 16 - 2 * 7 * 4 * 11/14 = 21$. Therefore, $AM = \sqrt{21}$.

The answer is (C).

Solution 2:

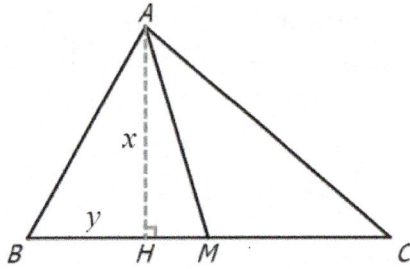

Let H be is a point on BC such that $AH \perp BC$. Letting $x = AH$ and $y = BH$, we get $HM = 4 - y$ and $HC = 8 - y$.

By the **Pythagorean Theorem**, for the right triangle ABH and ACH, we have $AB^2 = AH^2 + BH^2$ and $AC^2 = AH^2 + HC^2$. Thus $25 = x^2 + y^2$ and $49 = x^2 + (8 - y)^2$.

Solving the equation set gives $y = 5/2$ and $x = 5\sqrt{3}/2$. Therefore, $AM^2 = 21$ and $AM = \sqrt{21}$.

The answer is (C).

Solution 3:

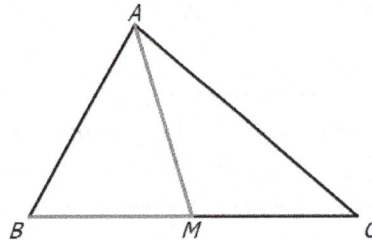

Using the **Median Length Formula** yields $AM^2 + BM^2 = (AB^2 + AC^2)/2$. Therefore, $AM^2 = (5^2 + 7^2)/2 - 4^2 = 21$, and $AM = \sqrt{21}$.

The answer is (C).

Problem 134

Two sides of a triangle have lengths 25 and 20, and the median to the third side has length 19.5. Find the length of the third side. ($AB = 20$, $AC = 25$ and $AM = 19.5$)

(A) 22.5   (B) 23   (C) 23.5   (D) 24   (E) 24.5

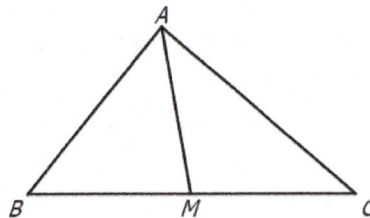

Tips: *Construct the altitude on side BC, use the law of cosines, or use the median length formula.*

Solution 1:

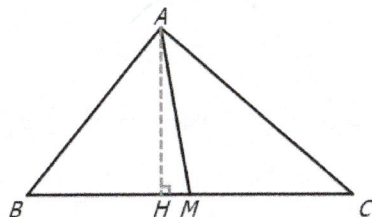

Let H be a point on BC such that $AH \perp BC$. Let $x = MC = MB$ and $y = HM$, so $BH = x - y$ and $HC = x + y$.

By the **Pythagorean Theorem**, for the three right triangle ABH, AHM and ACH, we have $AH^2 = AB^2 - BH^2 = AM^2 - HM^2 = AC^2 - HC^2$.

Thus $400 - (x - y)^2 = 19.5^2 - y^2$ and $19.5^2 - y2 = 625 - (x + y)^2$

Solving the equation set gives $x^2 = 132.25$. Therefore, $BC = 2x = 23$.

The answer is (B).

Solution 2:

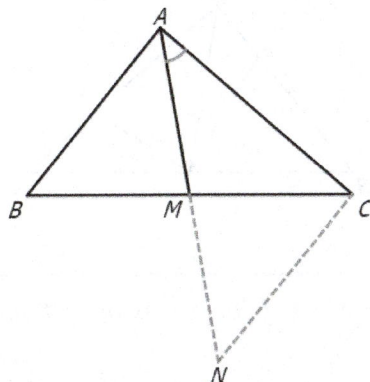

Extend AM to point N such that $MN = AM$ and connect NC. Because $BM = CM$ and $AM = MN$, $CN = AB = 4$ and $AN = 2AM = 6$.

Using the law of cosines in $\triangle ANC$, we have

$$cos(\angle CAM) = \frac{AN^2 + AC^2 - CN^2}{2 * AN * AC} = \frac{39^2 + 25^2 - 20^2}{2 \times 39 \times 25} = \frac{291}{325}$$

In the triangle AMC, $MC^2 = AM^2 + AC^2 - 2AM * AC * \cos\angle CAM = 19.5^2 + 25^2 - 2 * 19.5 * 25 * 291/325 = 132.25$

Therefore, $BC = 2MC = 23$.

The answer is (B).

Solution 3:

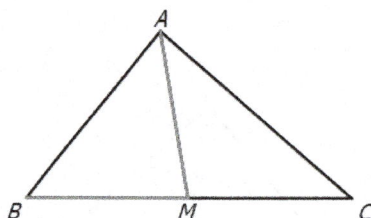

Using the **Median Length Formula** yields $AM^2 + BM^2 = (1/2) * (AB^2 + AC^2)$. Thus $BM^2 = (20^2 + 25^2)/2 - 19.5^2 = 132.25$. Therefore, $BC = 2BM = 23$.

The answer is (B).

Problem 135

In right triangle ABC, $\angle BAC = 90°$ and P, Q are the points on AB and AC such that $MP \perp MQ$. If $CQ = x$ and $BP = y$, what is $PM^2 + QM^2$?

(A) $2xy$ (B) $x^2 + y^2$ (C) $(x + y)^2$ (D) $x^2 + xy + y^2$

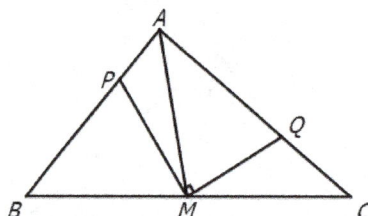

> **Tips:** *Use median to construct a congruent triangle and transport QC or PB to one triangle.*

Solution 1:

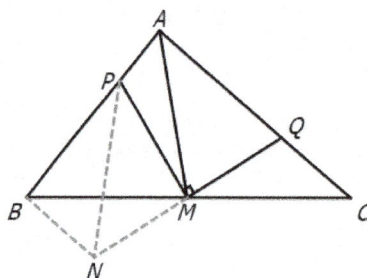

Extend QM to N such that $MN = QM$. Because $MN = QM$ and $BM = MC$, $\triangle CQM \cong \triangle BNM$. Therefore, $BN = CQ = x$ and $\angle MBN = \angle C$. So $BN//AC$.

Because $BN//AC$, $\angle NBP = \angle BAC = 90°$. Therefore, $PM^2 + QM^2 = PN^2 = BP^2 + BN^2 = BP^2 + CQ^2 = x^2 + y^2$.

The answer is (B).

Solution 2:

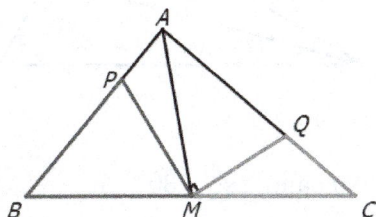

Let R_1 and R_2 be the radii of the circumcicles of triangle QMC and PMB, respectively. Applying the law of sines in $\triangle QMC$ and $\triangle PMB$ yields

$$2R_1 = \frac{MC}{\sin \angle MQC} = \frac{QM}{\sin \angle C} = \frac{CQ}{\sin \angle QMC}$$

and

$$2R_2 = \frac{MB}{\sin \angle MPB} = \frac{PM}{\sin \angle B} = \frac{BP}{\sin \angle PMB}$$

Because $\angle BAC = \angle PMQ = 90°$, $\angle APM + \angle AQM = 180°$ and $\angle BPM + \angle CQM = 180°$, we have $\sin \angle BPM = \sin \angle BPM$. Together with the fact that $MB = MC$, we obtain

$$2R_1 = 2R_2 = \frac{QM}{\sin \angle C} = \frac{CQ}{\sin \angle QMC} = \frac{MC}{\sin \angle MQC} = \frac{PM}{\sin \angle B} = \frac{BP}{\sin \angle PMB} = 2R$$

Because $\angle B + \angle C = 90°$ and $\angle QMC + \angle PMB = 90°$, $\sin \angle B = \cos \angle C$ and $\sin \angle PMB = \cos \angle MQC$. Therefore, $PM^2 + QM^2 = (2R)^2 = BP^2 + CQ^2 = x^2 + y^2$.

The answer is (B).

Problem 136

In the right $\triangle ABC$, $\angle C = 90°$ and $\angle A = 30°$. AE is the angular bisector of angle A. If $AE = 1$ cm, what is the area of triangle ABC?

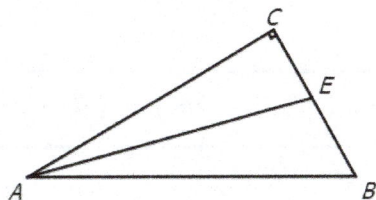

Tips: *Use the properties of a right triangle a 30° angle.*

Solution 1:

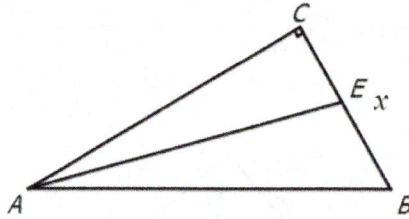

Let $x = BC$. Because $\angle C = 90°$ and $\angle A = 30°$, $AC = \sqrt{3}x$ and $AB = 2x$.

Because AE is the angular bisector of $\angle A$, $AB/AC = BE/CE = 2/\sqrt{3}$. Then, $BE = 2/(2 + \sqrt{3})x$ and $CE = \sqrt{3}/(2 + \sqrt{3})x$.

Applying the **Angular Bisector Length Formula** yields $AE^2 = AB * AC - BE * CE$, so $1 = 12(2 - \sqrt{3})x^2$ and $x^2 = (2 + \sqrt{3})/12$. Therefore, the area of triangle ABC is $\sqrt{3}x^2/2 = (3 + 2\sqrt{3})/24$.

Solution 2:

In the right triangle ACE, $\angle CAE = 15°$ and $AC = AE \cos \angle CAE = \cos 15°$.

Using the trigonometric identities gives $\cos 15° = \cos(45° - 30°) = \cos 45° \cos 30° + \sin 45° \sin 30° = (\sqrt{6} + \sqrt{2})/4$

Then, $AC = (\sqrt{6} + \sqrt{2})/4$ and the area of the triangle is $AC^2/(2\sqrt{3}) = (3 + 2\sqrt{3})/24$.

Problem 137

In the right $\triangle ABC$, CD is the altitude on side AB and AF is the angular bisector of $\angle BAC$. AF meets CD and CB at point E and F. G is a point on BC such that $EG//AB$. If $CE = \sqrt{2}$, find BG.

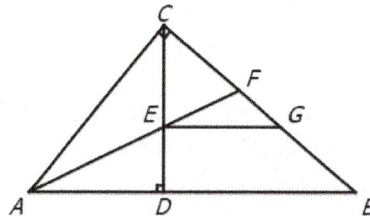

Tips: *Use the Angular Bisector Theorem and information from parallel lines.*

Solution 1:

There are three geometric features that involve ratio of segments.

Because AF is the angular bisector of $\angle CAB$, $AC/AD = CE/ED$. $EG//AB$ yields $CE/ED = CG/GB$. So $AC/AD = CG/GB$.

It is straightforward to see that $\triangle ACD \sim \triangle CGE$, so $AC/AD = CG/CE$.

Therefore, $CG/GB = CG/CE$ and $GB = CE = \sqrt{2}$.

Solution 2:

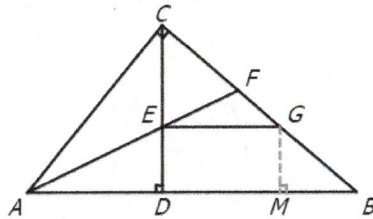

Let point M be on segment DB such that $GM \perp DB$.

Because AF is the angular bisector of $\angle CAB$, $AC/AD = CE/ED$.

Because $\triangle ACD \sim \triangle GBM$, $AC/AD = GB/GM$, so $CE/ED = GB/GM$.

Quadrilateral $DEGM$ is a rectangle and $ED = GM$. Therefore, $GB = CE = \sqrt{2}$.

Problem 138

In the right $\triangle ABC$, if $\angle A = 60°$, prove $AB = 2AC$.

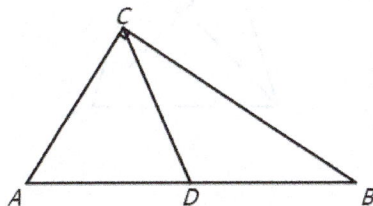

> **Tips:** *Create an equilateral triangle first.*

Solution:

Let point D be on AB such that $AD = AC$. Because $\angle A = 60°$ and $AD = AC$, triangle ACD is an equilateral triangle. Thus, $AC = AD = CD$ and $\angle ACD = 60°$.

$\angle ACB = 90°$ and $\angle ACD = 60°$ yields $\angle BCD = 30°$. Because $\angle BCD = \angle B = 30°$, $BD = CD = AC = AD$. Therefore, $AB = BD + AD = 2AC$.

213

Problem 139

Triangle AMC is an isosceles triangle with $AM = AC$. Medians MV and CU are perpendicular to each other, and $MV = CU = 12$. What is the area of $\triangle AMC$? (2020 AMC10A)

(A) 48   (B) 72   (C) 96   (D) 144   (E) 192

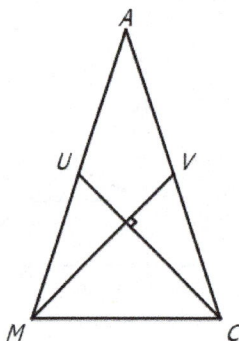

> **Tips:** *MV and CU intersects at centroid of triangle AMC, so use the properties of a centroid.*

Solution 1:

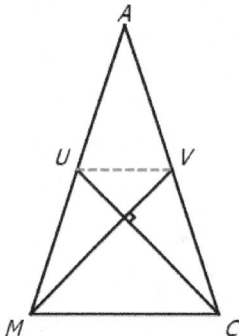

Connect UV, which is a midline of triangle AMC. Thus, the area of triangle AUV is 1/4 of the area of triangle AMC and the area of quadrilateral $UVCM$ is 3/4 of the area of triangle AMC.

Because $MV \perp UC$, the area of $UVCM$ is $UC * MV/2 = 12 \times 12/2 = 72$, so the area of triangle AMC is $72/(3/4) = 96$.

The answer is (C).

Solution 2:

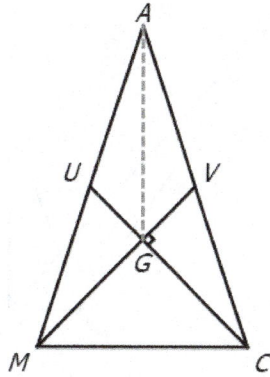

Let G be the intersection point of MV and UC. Because MV and CU are the medians of triangle AMC, G is the centroid of $\triangle AMC$ and $VG = MV/3$. $UG = CU/3$ and area of triangle GMC is $1/3$ of the area of triangle AMC.

Therefore, $CG = MG = 12 \times 2/3 = 8$ and the area of triangle GMC is $8 \times 8/2 = 32$, so the area of triangle AMC is $32 \times 3 = 96$.

The answer is (C).

Problem 140

Quadrilateral $ABCD$ satisfies angle $\angle ABC = \angle ACD = 90°$, $AC = 20$, and $CD = 30$. Diagonals AC and BD intersect at point E, and $AE = 5$. What is the area of quadrilateral $ABCD$? (2020 AMC10A)

(A) 330   (B) 340   (C) 350   (D) 360   (E) 370

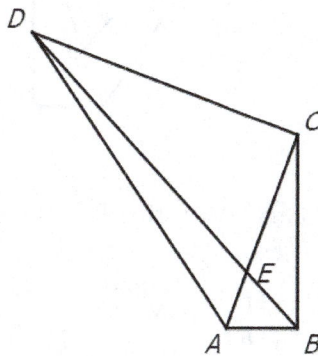

> **Tips:** *Calculate the length of the altitude to AE in triangle ABE.*

Solution 1:

215

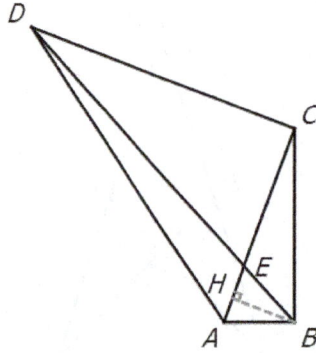

Let H be on segment AE such that $BH \perp AE$. Let $x = HE$. Because $BH \perp AE$ and $CD \perp AC$, $\triangle CDE \sim \triangle BEH$.

Because $\triangle CDE \sim \triangle BEH$, $HB/HE = CD/CE = 30/15 = 2$. So $HB = 2x$, $CH = 15 + x$ and $AH = 5 - x$. In the right triangle ABC, BH is the altitude. Therefore $BH^2 = AH * CH$ and $(2x)^2 = (5 - x) * (15 + x)$. Solving the equation yields $x = -5$ (ignore) and $x = 3$. So $BH = 2x = 6$.

The area of triangle ACD and ABC are 300 and 60, respectively, so the area of quadrilateral $ABCD$ is 360.

The answer is (D).

Solution 2:

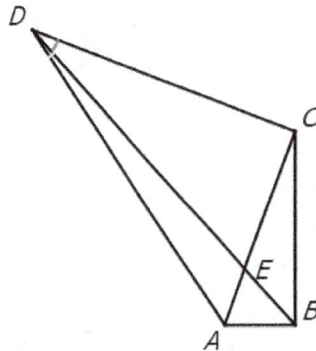

The area of triangle ACD is 300 and the ratio of the area of quadrilateral $ABCD$ to the area of triangle ACD is equal to BD/DE.

In right triangle ABD and CDE, $AD = 10\sqrt{13}$ and $DE = 15\sqrt{5}$. We also have $\cos \angle CDE = 2/\sqrt{5}$, $\sin \angle CDE = 1/\sqrt{5}$, $\cos \angle ADC = 3/\sqrt{13}$ and $\sin \angle ADC = 2/\sqrt{13}$. So $\cos \angle ADB = \cos(\angle ADC - \angle CDE) = 8/\sqrt{65}$.

Let $x = BD/DE$. By the law of cosines, $BC^2 = CD^2 + BD^2 - 2CD * BD * \cos \angle CDE$ and $AB^2 = AD^2 + BD^2 - 2AD * BD * \cos \angle ADC$. By the **Pythagorean Theorem**, $AC^2 = BC^2 + AB^2$. Therefore $400 = 900 + 1300 + 2250x^2 - 1800x - 2400x$. Simplifying the equation gives $15x^2 - 28x + 12 = 0$.

Solving the quadratic equation gives $x = 2/3$ (ignore, should be > 1) and $x = 6/5$. So the area of quadrilateral $ABCD$ is equal to $(6/5) * 300 = 360$.

The answer is (D).

Problem 141

A three-quarter sector of a circle of radius 4 inches along with its interior is the 2-D net that forms the lateral surface area of a right circular cone by taping together along the two radii shown. What is the volume of the cone in cubic inches? (2020 AMC10B)

(A) $3\pi\sqrt{5}$ (B) $4\pi\sqrt{3}$ (C) $3\pi\sqrt{7}$ (D) $6\pi\sqrt{3}$ (E) $6\pi\sqrt{7}$

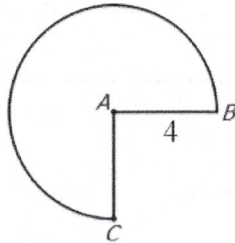

Tips: *The perimeter of the base circle of the cone is equal to the original length of arc BC.*

Solution:

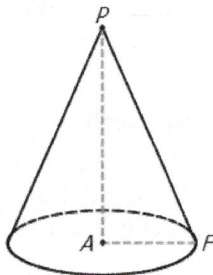

The length of the perimeter of the base circle is equation 3/4 of original circle perimeter, so $2\pi * AF = (3/4) * 2\pi * AB = (3/4) * AB = 3$ inches.

In the right triangle PAF, $PF = AB = 4$ inches, $AF = 3$ inches, so $PA = \sqrt{7}$.

The volume of the cone is $(1/3)\pi * AF^2 * PA = (1/3) * \pi * 9\sqrt{7} = 3\pi\sqrt{7}$.

The answer is (C).

Problem 142

As shown in the figure below, six semicircles lie in the interior of a regular hexagon with side length 2 so that the diameters of the semicircles coincide with the sides of the hexagon. What is the area of the shaded region — inside the hexagon but outside all of the semicircles? (2020 AMC10B)

(A) $6\sqrt{3} - 3\pi$ (B) $9\sqrt{3}/2 - 2\pi$ (C) $3\sqrt{3}/2 - \pi/3$ (D) $3\sqrt{3} - \pi$ (E) $9\sqrt{3}/2 - \pi$

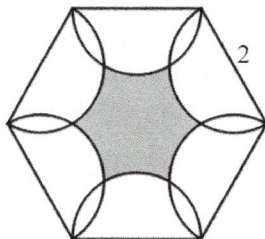

Tips: *Connect the centers of the circles and the interception points to the center of the hexagon.*

Solution:

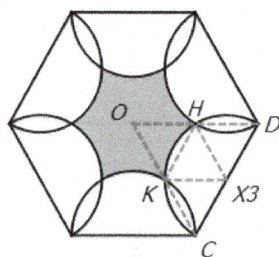

Connect the center O and two adjacent vertices C and D to form an equilateral triangle OCD (isosceles triangle OCD with $\angle ODC = 60°$). Connect the intersecting point H and center of the semicircle X_3 to form an smaller equilateral triangle DHX_3. The total shaded area in the triangle OCD is equal to the area difference of 12 small equilateral triangles and one full circle.

We know $OH = OK = CX_3 = 1$, so the area of two small equilateral triangles is $3\sqrt{3}$. Therefore, the shaded area is $3\sqrt{3} - \pi$.

The answer is (D).

Problem 143

In square $ABCD$, points E and H lie on AB and DA, respectively, so that $AE = AH$. Points F and G lie on BC and CD, respectively, and points I and J lie on EH so that $FI \perp EH$ and $GJ \perp EH$. See the figure below. Triangle AEH, quadrilateral $BFIE$, quadrilateral $DHJG$, and pentagon $FCGJI$ each has area 1. What is FI^2? (2020 AMC10B)

(A) $7/3$ (B) $8 - 4\sqrt{2}$ (C) $1 + \sqrt{2}$ (D) $7\sqrt{2}/4$ (E) $2\sqrt{2}$

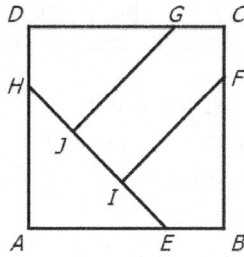

Tips: *1. Find the edge-length of the square. 2. Find the constraints involving FI in either CGJIF or FIEB.*

Solution 1:

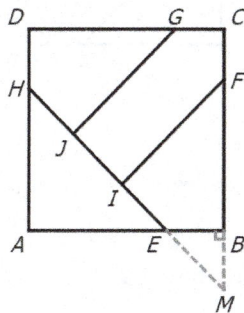

The total area of the square is 4 and the edge-length is 2. In the isosceles right triangle AEH, $AE = \sqrt{2}$ because its area is 1, and then $BE = 2 - \sqrt{2}$.

Extend HE and CB to meet at point M. Then, both ΔFIM and ΔEBM are isosceles right triangles.

The area of ΔEBM is equal to $BE^2/2 = 3 - 2\sqrt{2}$ and the area of ΔFIM is equal to $FI^2/2 = 3 - 2\sqrt{2} + 1 = 4 - 2\sqrt{2}$. Therefore, $FI^2 = 8 - 4\sqrt{2}$.

The answer is (B).

Solution 2:

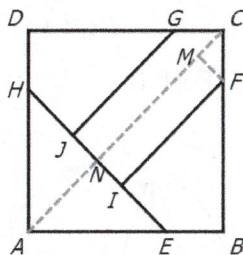

Connect AC and label the intersection with HE as point N. M is on AC such that $FM \perp AC$. The area of trapezoid $CNIF$ is then 0.5.

219

The total area of the square is 4 and the edge-length is 2. In the isosceles right triangle AEH, $AE = \sqrt{2}$ because its area is 1 and then $NE = 1$, $NC = 2\sqrt{2} - 1$.

Let $x = FI$. Then, the altitude of trapezoid $CNIF$ is $MF = NC - FI = 2\sqrt{2} - 1 - x$ and the area of trapezoid $CNIF$ is $(2\sqrt{2} - 1 + x)(2\sqrt{2} - 1 - x)/2 = 0.5$. Simplifying the equation yields $x^2 = (2\sqrt{2} - 1)^2 - 1 = 8 - 4\sqrt{2}$.

The answer is (B).

Solution 3:

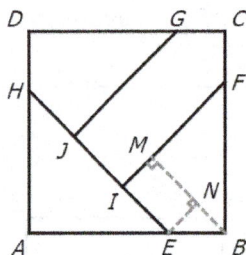

Let M be a point on FI such that $MB \perp FI$ and N is a point on MB such that $EN \perp MB$. Both $\triangle FMB$ and $\triangle ENB$ are isosceles right triangles.

The total area of the square is 4 and the edge-length is 2. In the isosceles right triangle AEH, $AE = \sqrt{2}$ because its area is 1 and then $BE = 2 - \sqrt{2}$. So $MI = EN = NB = \sqrt{2} - 1$.

Let $x = FI$. In the trapezoid $MIEB$, we have $MB = x - MI$, $IE = MB - NB = x - 2MI$. Then, area of trapezoid $MIEB$ is equal to $(2x - 3MI) * MI/2$. and the area of $FIEB$ is $(x - MI)^2/2 + (2x - 3MI) * MI/2 = 1$. Simplifying the equation gives $x^2 = 2MI^2 + 2 = 2(\sqrt{2} - 1)2 + 2 = 8 - 4\sqrt{2}$.

The answer is (B).

Problem 144

Let $\triangle ABC$ be an isosceles triangle with $BC = AC$ and $\angle ACB = 40°$. Construct the circle with diameter BC, and let D and E be the other intersection points of the circle with the sides AC and AB, respectively. Let F be the intersection of the diagonals of the quadrilateral $BCDE$. What is the degree measure of $\angle BFC$? (2019 AMC10A)

(A) 90   (B) 100   (C) 105   (D) 110   (E) 120

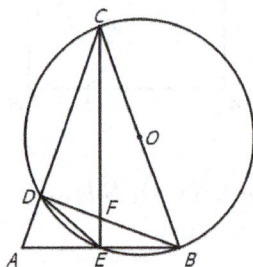

Tips: *Use the angle properties of a concyclic quadrilateral.*

Solution 1:

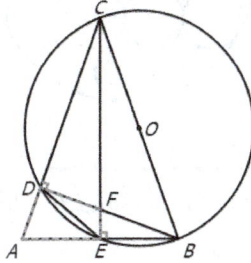

Because $CA = CB$ and $\angle ACB = 40°$, $\angle A = (180° - 40°)/2 = 70°$.

Because BC is the diameter of the circle, $\angle CDB = \angle CEB = 90°$. Hence points A, D, F and E are concyclic, which yields $\angle A + \angle DFE = 180°$. Thus, $\angle BFC = \angle DFE = 180° - \angle A = 180° - 70° = 110°$.

The answer is (D).

Solution 2:

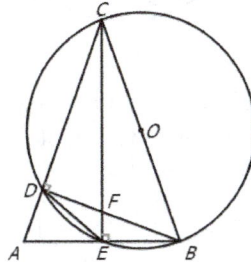

BC is the diameter of the circle, so $\angle CDB = \angle CEB = 90°$. Because $CA = CB$ and CE is the median, CE is the angular bisector to side AB. Therefore, $\angle BCE = \angle ACE = 40°/2 = 20°$.

Because the points B, C, D, E are on the circle, $\angle DBE = \angle ACE = 20°$. Thus, $\angle BFC = \angle DBE + \angle FEB = 20° + 90° = 110°$.

The answer is (D)

Problem 145

The figure below shows 13 circles of radius 1 within a larger circle. All the intersections occur at points of tangency. What is the area of the region, shaded in the figure, inside the larger circle but outside all the circles of radius 1? (2019 AMC10A)

(A) $4\pi\sqrt{3}$ (B) 7π (C) $(3\sqrt{3}+2)\pi$ (D) $10(\sqrt{3}-1)\pi$ (E) $(\sqrt{3}+6)\pi$

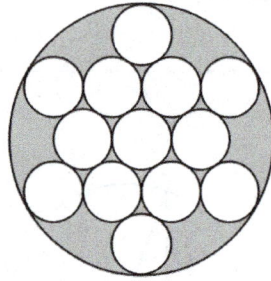

Tips: *Connect the centers of the circles and the tangent points to find the radius of the large circle.*

Solution:

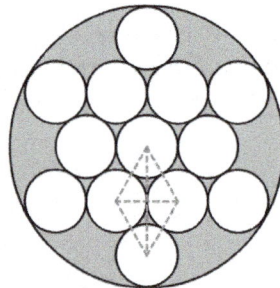

Connecting the centers of the circles as shown in the diagram gives us two equilateral triangles. Given the radius of the small circles is 1, the altitude of the equilateral triangle is $\sqrt{3}$. The radius of the big circle is $\sqrt{3} * 2 + 1 = 2\sqrt{3} + 1$.

Thus, the area of shaded region is $\pi(2\sqrt{3} + 1)^2 - 13\pi = 4\pi\sqrt{3}$.

The answer is (A).

Problem 146

A sphere with center O has radius 6. A triangle with sides of length 15, 15, and 24 is situated in space so that each of its sides are tangent to the sphere. What is the distance between O and the plane determined by the triangle? (2019 AMC10A)

(A) $2\sqrt{3}$ (B) 4   (C) $3\sqrt{2}$ (D) $2\sqrt{5}$ (E) 5

Tips: *1. Find the radius of the inscribed circle of the triangle. 2. Construct a right triangle with the center of the sphere, the center of the inscribed circle, and the tangent point as vertices.*

Solution:

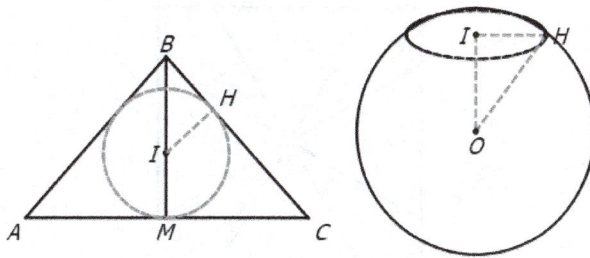

The tangent points between the triangle and sphere are the three points of the inscribed circle of triangle as shown in the diagram left. In the triangle, $BA = BC = 15$ and $AC = 24$, so $MC = MA = 12$, $BM = 9$ and $BH = 15 - 12 = 3$.

Let the radius of the inscribed circle be r. Then, $(9-r)^2 = r^2 + 9$. Solving the equation yields $HI = r = 4$.

Connect the centers of the sphere (O) and the inscribed circle (I). In the right triangle OHI, the distance $OI = \sqrt{OH^2 - HI^2} = 2\sqrt{5}$.

The answer is (D).

Problem 147

The figure below shows a square and four equilateral triangles, with each triangle having a side lying on a side of the square, such that each triangle has side length 2 and the third vertices of the triangles meet at the center of the square. The region inside the square but outside the triangles is shaded. What is the area of the shaded region? (2019 AMC10B)

(A) 4   (B) $12 - 4\sqrt{3}$ (C) $3\sqrt{3}$ (D) $4\sqrt{3}$ (E) $16 - 4\sqrt{3}$

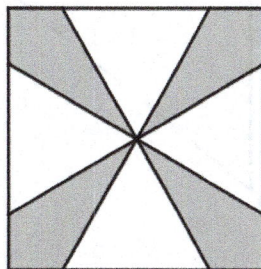

Tips: *The edge-length of the square is twice as long as the altitude of the equilateral triangles.*

Solution:

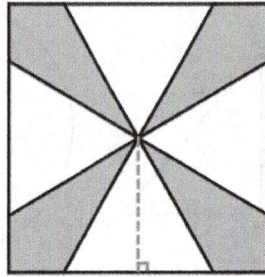

The altitude of the equilateral triangle is $(\sqrt{3}/2) * 2 = \sqrt{3}$. Thus, the edge-length of the square is $2\sqrt{3}$. The area of the square is 12 and the total area of four equilateral triangles is $(\sqrt{3}/4) * 2^2 * 4 = 4\sqrt{3}$.

Therefore, the area of the shaded region is $12 - 4\sqrt{3}$.

The answer is (B).

Problem 148

Right triangles T_1 and T_2 have areas 1 and 2, respectively. A side of T_1 is congruent to a side of T_2, and a different side of T_1 is congruent to a different side of T_2. What is the square of the product of the other (third) sides of T_1 and T_2? (2019 AMC10B)

(A) 28/3   (B) 10   (C) 32/3   (D) 34/3   (E) 12

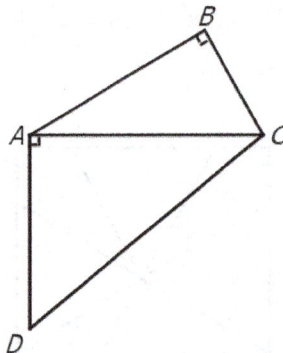

Tips: *Based on the ratio of side lengths, one of the right triangles has a special angle.*

Solution 1:

224

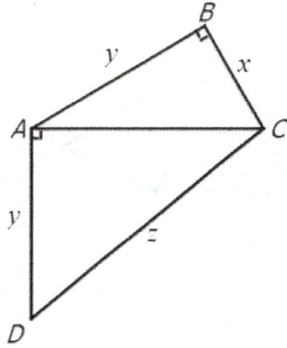

Let $x = BC$, $AD = AB = y$ and $z = CD$. Then, $AC = \sqrt{x^2 + y^2}$. Given the area of triangles ABC and ACD as 1 and 2, we have $xy = 2$ and $y\sqrt{x^2 + y^2} = 4$. So $x^2y^2 = 4$ and $y^4 = 12$ and $x^4 = 16/12 = 4/3$.

Therefore, $x^2z^2 = x^2(x^2 + 2y^2) = x^4 + 2x^2y^2 = 4/3 + 8 = 28/3$.

The answer is (A).

Solution 2:

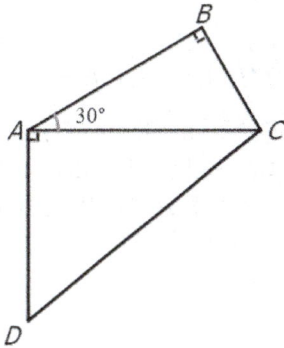

The area of the triangles ABC and ADC are $AB*BC/2 = 1$ and $AD*AC/2 = 2$, respectively. Knowing $AB = AD$, we get $AC/BC = 2$. Because $AC = 2BC$ in the right triangle, $\angle BAC = 30°$ and $AB = \sqrt{3}BC$. So $BC^2 = 2/\sqrt{3}$, $AB^2 = 2\sqrt{3}$, and $AC^2 = 2(\sqrt{3} + 1/\sqrt{3}) = (8/3)*\sqrt{3}$.

In the right triangle ACD, $CD^2 = AD^2 + AC^2 = AB^2 + AC^2 = 2\sqrt{3} + (8/3)*\sqrt{3} = (14/3)*\sqrt{3}$. Thus, $BC^2 * CD^2 = (14/3) * \sqrt{3} * (2/\sqrt{3}) = 28/3$.

The answer is (A).

Problem 149

In $\triangle ABC$ with a right angle at C, point D lies in the interior of AB and point E lies in the interior of BC so that $AC = CD$, $DE = EB$, and the ratio $AC : DE = 4 : 3$. What is the ratio $AD : DB$? (2019 AMC10B)

(A) $2 : 3$ (B) $2 : \sqrt{5}$ (C) $1 : 1$ (D) $3 : \sqrt{5}$ (E) $3 : 2$

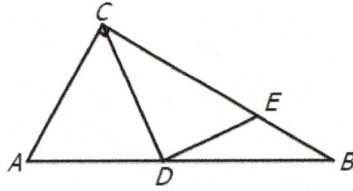

Tips: *Prove triangle CDE is a right triangle.*

Solution 1:

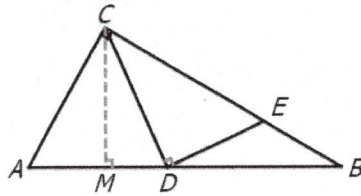

Because $AC = CD$ and $DE = DE$, $\angle A = \angle CDA$ and $\angle B = \angle EDB$. In the right triangle ABC, $\angle A + \angle B = 90°$. Thus, $\angle CDA + \angle EDB = 90°$ and $\angle CDE = 90°$.

Let $AC = CD = 4$ and $EB = ED = 3$. Then, $CE = 5$, $BC = 5 + 3 = 8$, and $AB = 4\sqrt{5}$. Because CM is the altitude of the right triangle, $AC^2 = AB * AM$. Solving the equation gives us $AM = 4/\sqrt{5}$. $AD/AB = 2AM/AB = 2/5$. Therefore, $AD/DB = 2/3$.

The answer is (A).

Solution 2:

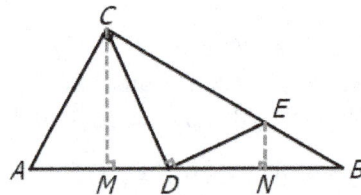

In this solution, we will solve the problem using trigonometry. Because $AC = CD$ and $DE = DE$, $\angle A = \angle CDA$ and $\angle B = \angle EDB$. In the right triangle ABC, $\angle A + \angle B = 90°$. We have $\triangle CDM \sim \triangle DEN$ and $MD = MA$, $NB = ND$.

Let $\angle EDB = \angle B = \angle DCM = \beta$, and $CA = CD = 4$ and $ED = EB = 3$. So $MA = MD = 4\sin\beta$, $CM = 4\cos\beta$, $BD = 2ED\cos\beta = 6cos\beta$.

In the right triangle ABC, CM is the altitude of side AB. Thus, $CM^2 = MA * MB$, which yields $16(\cos\beta)^2 = 4\sin\beta(4\sin\beta + 6\cos\beta)$. Simplifying the equation yields $16(\cos(2\beta)) = 12\sin(2\beta)$ and $\tan(2\beta) = 4/3$. By the double angle identity of tangent function, we have $2\tan\beta/(1 - (tan\beta)^2) = 4/3$. Solving the equation gives $\tan\beta = 1/2$ or $\tan\beta = -2$ (ignore).

Therefore, $AD : DB = (4\sin\beta)/(3\cos\beta) = (4/3) * \tan\beta = (4/3) \times (1/2) = 2 : 3$.

The answer is (A).

Problem 150

As shown in the figure, line segment AD is trisected by points B and C so that $AB = BC = CD = 2$. Three semicircles of radius 1, AEB, BFC, and CGD, have their diameters on AD, and are tangent to line EG at E, F, and G, respectively. A circle of radius 2 has its center on F. The area of the region inside the circle but outside the three semicircles, shaded in the figure, can be expressed in the form $\frac{a}{b}\pi - \sqrt{c} + d$, where a, b, c, and d are positive integers and a and b are relatively prime. What is $a + b + c + d$? (2019 AMC10B)

(A) 13   (B) 14   (C) 15   (D) 16   (E) 17

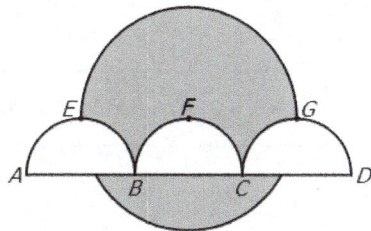

> **Tips:** *Connect the centers of the circles to the intersection points to form right triangles.*

Solution:

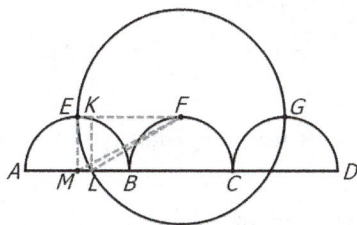

Label the points as shown in the diagram. Let the area of the big circle be S_{O1}, the area of one small circle be S_{O2}, and the area enclosed by EML be S_3. The area of the shade region is equal to $S_{O1} - S_{O2} + 2S_3$.

Construct the diagram as shown above. $KL = EM = 1 = FL/2$. Thus in the right triangle KLF, $\angle EFL = 30°$ and the area of sliced pie FEL in the big circle is $4\pi/12 = \pi/3$.

In the trapezoid $EFLM$, $ML = EK = 2 - \sqrt{3}$, so the area of the trapezoid is equal to $(2 + 2 - \sqrt{3}) * (1/2) = (4 - \sqrt{3})/2$. Therefore, the small area $S_3 = (4 - \sqrt{3})/2 - \pi/3$.

The area of the shaded region is equal to $(4 - 1)\pi + 2((4 - \sqrt{3})/2 - \pi/3) = \frac{7}{3}\pi - \sqrt{3} + 4$, so $a + b + c + d = 17$.

The answer is (E).

227

Problem 151

All of the triangles in the diagram below are similar to isosceles triangle ABC, in which $AB = AC$. Each of the 7 smallest triangles has area 1, and $\triangle ABC$ has area 40. What is the area of trapezoid $DBCE$? (2018 AMC10A)

(A) 16   (B) 18   (C) 20   (D) 22   (E) 24

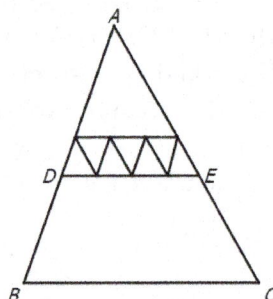

Tips: *1. All of the small triangles are similar to the large triangle.*

Solution 1:

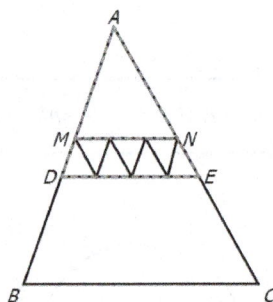

Because all of the small triangles are congruent, $MN/DE = 3/4$. Thus $S_{\triangle AMN}/S_{\triangle ADE} = (3/4)^2 = 9/16$.

Given that the area of $MNDE$ is 7, we know $S_{\triangle ADF} = 16$. Therefore, the area of trapezoid $DECB$ is equal to $40 - 16 = 24$.

The answer is (E).

Solution 2:

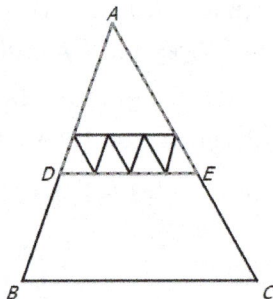

228

Because the small triangles are similar to the big triangle and their areas are 1 and 40, respectively, the ratio of their edge-lengths is $1/\sqrt{40}$.

Thus, $DE/BC = 4/\sqrt{40}$. Given $\triangle ADE \sim \triangle ABC$, we have $S_{\triangle ADE}/S_{\triangle ABC} = (4/\sqrt{40})^2 = 16/40$. Therefore, $S_{\triangle ADE} = 16$ and the area of trapezoid $DEBC$ is equal to $40 - 16 = 24$.

The answer is (E).

Solution 3:

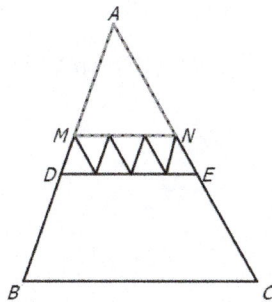

Similar to solution 2, we find that the ratio of the edge-length of the small triangles to the large triangle is $1/\sqrt{40}$.

Thus $MN/BC = 3/\sqrt{40}$. Given $\triangle AMN \sim \triangle ABC$, we have $S_{\triangle AMN}/S_{\triangle ABC} = (3/\sqrt{40})^2 = 9/40$. Therefore, $S_{\triangle AMN} = 9$ and the area of trapezoid $DECB$ is equal to $40 - 9 - 7 = 24$.

The answer is (E).

Problem 152

A paper triangle with sides of lengths 3, 4, and 5 inches, as shown, is folded so that point A falls on point B. What is the length in inches of the crease? (2018 AMC10A)

(A) $1 + \sqrt{2}/2$　　(B) $18\sqrt{3}$　　(C) $7/4$　　(D) $15/8$　　(E) 2

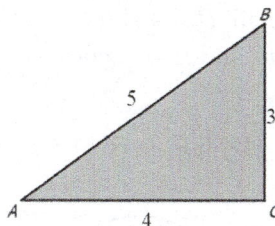

> **Tips:** *1. The crease is along the perpendicular bisector of segment AB.*

Solution 1:

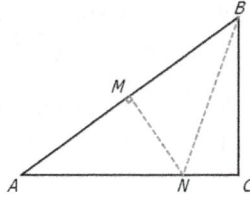

The crease can be represented by the perpendicular bisector MN on the hypotenuse AB, so $AM = 5/2$.

Because $\triangle AMN \sim \triangle ACB$, $MN/BC = AM/AC$. Thus $MN = (AM/AC) * BC = (5/2)/4 * 3 = 15/8$.

The answer is (D).

Solution 2:

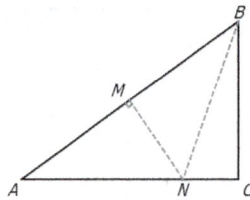

Let $x = NC$. In the right triangle AMN and BCN, by the **Pythagorean Theorem**, $4 - x = \sqrt{3^2 + x^2}$. Solving the equation yields $x = 7/8$.

Therefore, $MN = \sqrt{AN^2 - AM^2} = \sqrt{(4 - 7/8)^2 - (5/2)^2} = 15/8$.

The answer is (D).

Problem 153

Two circles of radius 5 are externally tangent to each other and are internally tangent to a circle of radius 13 at points A and B, as shown in the diagram. The distance AB can be written in the form m/n, where m and n are relatively prime positive integers. What is $m+n$? (2018 AMC10A)

(A) 21   (B) 29   (C) 58   (D) 69   (E) 93

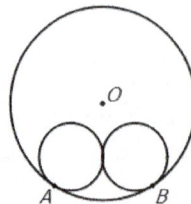

Tips: *1. Connect the centers of the circles to the tangent points. 2. Use properties of similar triangles.*

Solution 1:

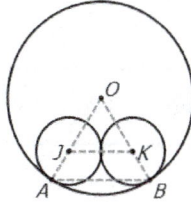

Let points J and K be the center of the two small circles. Connect OA, OB, AB and JK to form the similar triangle pair $\triangle OJK \sim \triangle OAB$.

It is straightforward to find that $JK = 10$, $OB = OA = 13$, and $OJ = OK = 13 - 5 = 8$.

Because $\triangle OJK \sim \triangle OAB$, $AB/JK = OB/OK$. Thus $AB = JK * OB/OK = 10 \times 13/8 = 65/4$. Therefore, $m = n = 65 + 4 = 69$.

The answer is (D).

Solution 2:

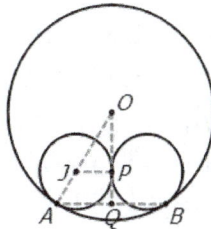

Connect OA, OP and JP, in which J is the center of a small circle and P is the tangent point of the two small circles. Then we have $OA = 13$, $JA = 5$, $OJ = 13 - 5 = 8$, and $JP = 5$.

Because $\triangle OJP \sim \triangle OAQ$, $AQ = OA/OJ * JP = 13/8 * 5 = 65/8$ and $AB = 2AQ = 65/4$. Therefore $m + n = 65 + 4 = 69$.

The answer is (D).

Problem 154

Farmer Pythagoras has a field in the shape of a right triangle. The right triangle's legs have lengths 3 and 4 units. In the corner where those sides meet at a right angle, he leaves a small unplanted square S so that from the air it looks like the right angle symbol. The rest of the field is planted. The shortest distance from S to the hypotenuse is 2 units. What fraction of the field is planted? (2018 AMC10A)

(A) 25/27   (B) 26/27   (C) 73/75   (D) 145/147   (E) 74/75

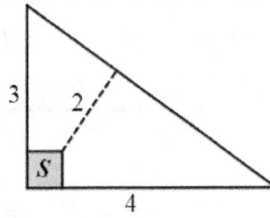

Tips: *Construct similar right triangles and use the Pythagorean Theorem.*

Solution 1:

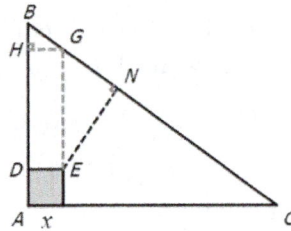

Let points H and G be on AB and BC such that $GE//AB$ and $HG//AC$. Thus $DEGH$ is a rectangle. In the right triangle ABC, $AB = 3$ and $AC = 4$, so $BC = 5$.

Let $x = AD$. We have $\triangle BHG \sim \triangle GNE \sim \triangle ABC$. Thus $HD = GE = (5/4) * EN = (5/4) \times 2 = 5/2$ and $BH = 3 - 5/2 - x = 1/2 - x$.

In the right triangle BHG, $BH/HG = 3/4$. Therefore $(1/2 - x)/x = 3/4$. Solving the equation yields $x = 2/7$. The total area is $3 \times 4/2 = 6$ and the planted area is $6 - 4/49$, so the ratio of planted area to total area is $145/147$.

The answer is (D).

Solution 2:

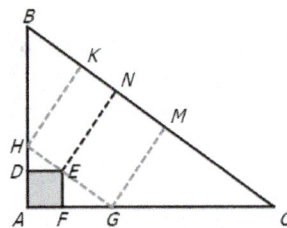

Construct a parallel line HG through point E such that $HG//BC$ and points H and G are on side AB and AC, respectively. The points K and M are on the hypotenuse such that $HK \perp BC$ and $GM \perp BC$. Thus $HGMK$ is a rectangle and $HK = GM = EN = 2$. In the right triangle ABC, $AB = 3$ and $AC = 4$ yields $BC = 5$.

We have $\triangle BHK \sim \triangle GCM \sim \triangle ABC$. Thus $BH = 5/4HK = 5/4 \times 2 = 5/2$ and $GC = (5/3) * GM = 5/3 \times 2 = 10/3$. Therefore, $HA = 3 - 5/2 = 1/2$ and $AG = 4 - GC = 2/3$.

Let $x = AD$. Because $\triangle HDE \sim \triangle EFG$, $HD/EF = DE/FG$ and then $(1/2 - x)/x = x/(2/3 - x)$. Solving the equation yields $x = 2/7$. The total area is $3 \times 4/2 = 6$ and the planted area is $6 - 4/49$, so the ratio of planted area to total area is $145/147$.

The answer is (D).

Solution 3:

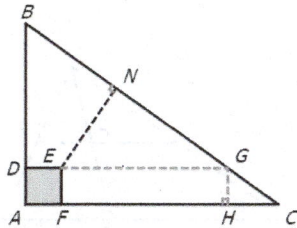

Similar to solution 1, the points H and G are on AC and BC such that $GE//AC$ and $HG//AB$, thus making $DEGH$ a rectangle. In the right triangle ABC, $AB = 3$ and $AC = 4$ yields $BC = 5$.

Let $x = AD$. We have $\triangle CHG \sim \triangle GNE \sim \triangle ABC$. Thus $FH = GE = 5/3 EN = 5/3 \times 2 = 10/3$ and $CH = 4 - 10/3 - x = 2/3 - x$.

In the right triangle CHG, $CH/HG = 4/3$. Therefore, we get the equation $(2/3 - x)/x = 4/3$. Solving the equation yields $x = 2/7$. The total area is $3 \times 4/2 = 6$ and the planted area is $6 - 4/49$. Therefore the ratio of planted area to total area is $145/147$.

The answer is (D).

Solution 4:

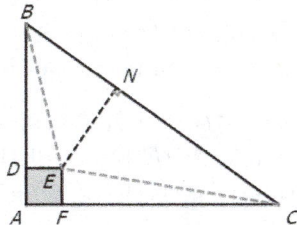

By the **Pythagorean Theorem** we have $BE^2 = DE^2 + BD^2 = EN^2 + BN^2$ and $CE^2 = EF^2 + FC^2 = EN^2 + NC^2$. Let $x = AD$, $BN = 5/2 - t$ and $CN = 5/2 + t$. We have

$$x^2 + (3 - x)^2 = 4 + (5/2 - t)^2 \quad \text{(a)}$$
$$x^2 + (4 - x)^2 = 4 + (5/2 + t)^2 \quad \text{(b)}$$

It may take some effort to solve this equation set. (b) - (a) yields $x = 7/2 - 5t$. Substituting it into equation (b) yields $49t^2 - 35t + 9/4 = 0$. Solving the quadratic equation gives $t = 9/14$ and $x = 2/7$ or $t = 1/14$ and $x = 22/7$ (ignore, should be < 3).

The total area is $3 \times 4/2 = 6$ and the planted area is $6 - 4/49$, so the ratio of planted area to total area is $145/147$.

The answer is (D).

233

Problem 155

Triangle ABC with $AB = 50$ and $AC = 10$ has an area of 120. Let D be the midpoint of AB, and let E be the midpoint of AC. The angle bisector of $\angle BAC$ intersects DE and BC at F and G, respectively. What is the area of quadrilateral $FDBG$? (2018 AMC10A)

(A) 60   (B) 65   (C) 70   (D) 75   (E) 80

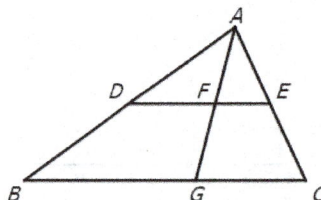

Tips: *Apply the angular bisector theorem.*

Solution 1:

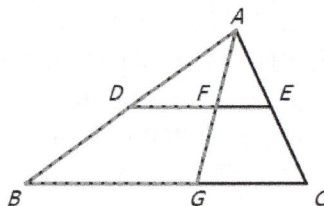

Because AG is the angular bisector of $\angle BAC$, the angular bisector theorem states $BG/GC = AB/AC = 50/10$. The ratio of the area of triangle ABG to the area of triangle AGC is equal to $BG/GC = 5$. Therefore, the area of triangle ABG is 100.

Because D and E are midpoints of AB and AC, respectively, DE is the midline of triangle ABC. Therefore, the area of triangle ADF is $1/4$ of the area of triangle ABG and the area of $FDBG$ is $3/4$ of the area of the triangle ABG, which is $100 \times 3/4 = 75$.

The answer is (D).

Solution 2:

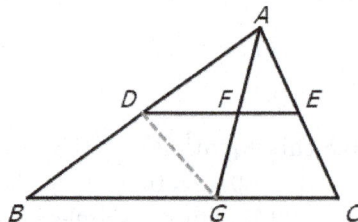

The areas of triangle ABG and triangle ACG are $S_{\triangle ABG} = (1/2) * AB * AG * \sin \angle BAG$ and $S_{\triangle ACG} = (1/2) * AC * AG * \sin \angle CAG$. Thus $S_{\triangle ABG}/S_{\triangle ACG} = AB/AC = 5$. Given the area of triangle ABC as 120, we know $S_{\triangle ABG} = 100$.

Because $AD = BD$ and $AF = FG$, $S_{\triangle BDG} = S_{\triangle ADG} = (1/2)S_{\triangle ABG} = 50$ and $S_{\triangle DFG} = (1/2)S_{\triangle ADG} = 25$. Therefore, the area of quadrilateral $FDBG$ is equal to $50 + 25 = 75$.

The answer is (D).

Problem 156

Line segment AC is a diameter of a circle with $AC = 24$. Point B, not equal to A or C, lies on the circle. As point B moves around the circle, the centroid (center of mass) of $\triangle ABC$ traces out a closed curve missing two points. To the nearest positive integer, what is the area of the region bounded by this curve? (2018 AMC10B)

(A) 25   (B) 38   (C) 50   (D) 63   (E) 75

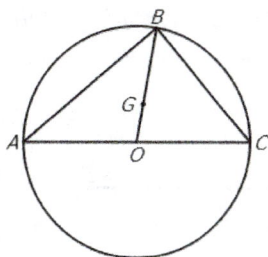

> **Tips:** *Remember that the centroid is the point of intersection of the medians.*

Solution:

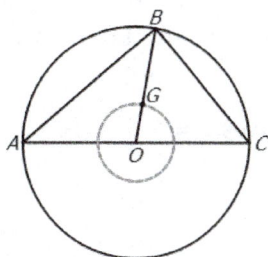

In the circle with AC as the diameter, $OA = OB = OC = 12$. Because G is the centroid of triangle ABC, $OG = (1/3) * OB = (1/3) * 12 = 4$. Thus, the curve forms a circle with a fixed radius of 4. Therefore, the area of the region bounded by this curve is $16\pi = 50.27 \approx 50$.

The answer is (C).

Problem 157

In rectangle $PQRS$, $PQ = 8$ and $QR = 6$. Points A and B lie on PQ, points C and D lie on QR, points E and F lie on RS, and points G and H lie on SP so that $AP = BQ < 4$ and the convex octagon $ABCDEFGH$ is equilateral. The length of a side of this octagon can be expressed in the form $k + m\sqrt{n}$, where k, m, and n are integers and n is not divisible by the square of any prime. What is $k + m + n$? (2018 AMC10B)

(A) 1   (B) 7   (C) 21   (D) 92   (E) 106

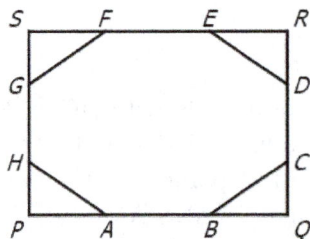

Tips: *Establish an equation using the Pythagorean Theorem of a corner right triangle.*

Solution 1:

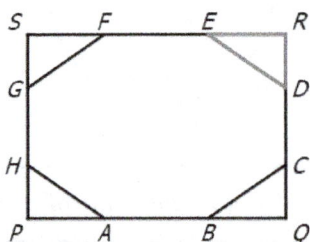

Let the side-length of the octagon be x. In the right triangle ERD, $ER = (8 - x)/2$ and $RD = (6 - x)/2$. By the **Pythagorean Theorem**, $DE^2 = ER^2 + RD^2$ and $x^2 = ((8 - x)/2)^2 + ((6 - x)/2)^2$.

Solving the quadratic equation gives $x = -7 + 3\sqrt{11}$. So $k + m + n = -7 + 3 + 11 = 7$.

The answer is (B).

Solution 2:

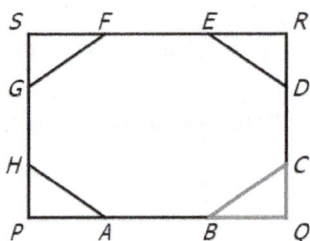

Let $x = PA = BQ$. The side-length of the octagon is equal to $(8 - 2x)$. Thus $CQ = (6 - (8 - 2x))/2 = x - 1$. By the **Pythagorean Theorem**, $BC^2 = CQ^2 + BQ^2$ and $(8 - 2x)^2 = x^2 + (x - 1)^2$.

236

Solving the equation gives $x = (15 + 3\sqrt{11})/2$. The side-length of the octagon is $8 - 2x = -7 + 3\sqrt{11}$, so $k + m + n = -7 + 3 + 11 = 7$.

The answer is (B).

Problem 158

Let $ABCDEF$ be a regular hexagon with side length 1. Let X, Y, and Z denote the midpoints of sides AB, CD, and EF, respectively. What is the area of the convex hexagon whose interior is the intersection of the interiors of $\triangle ACE$ and $\triangle XYZ$? (2018 AMC10B)

(A) $3/8\sqrt{3}$ (B) $7/16\sqrt{3}$ (C) $15/32\sqrt{3}$ (D) $1/2\sqrt{3}$ (E) $9/16\sqrt{3}$

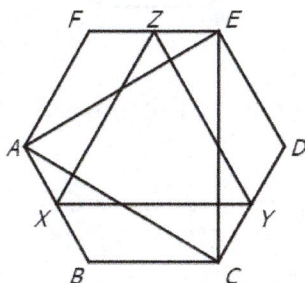

Tips: *The area is the difference between the areas of an equilateral triangle and the three small 30-60-90 right triangles.*

Solution 1:

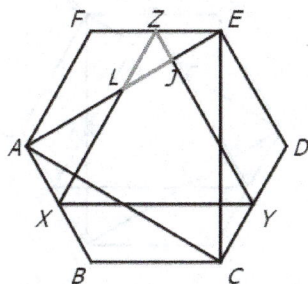

To find the targeted area, subtract the area of three right triangles ZLJ from the equilateral triangle XYZ.

In the right triangle ZLJ, $ZJ = (1/2) * ZE = (1/4) * FE = 1/4$. Thus $ZY = DE + 2ZJ = 1 + (1/4) * 2 = 3/2$. The area of equilateral triangle XYZ is $(\sqrt{3}/4) * ZY^2 = (\sqrt{3}/4) * (3/2)^2 = 9\sqrt{3}/16$.

In the right triangle ZLJ, $\angle ZLJ = 30°$ and $ZJ = 1/4$. Thus the area of $\triangle ZLJ$ is equal to $(1/2) * \sqrt{3} * (ZJ)^2 = \sqrt{3}/32$.

Extend HE and CB to meet at point M. Then both $\triangle FIM$ and $\triangle EFD$ are isosceles right triangles.

The area of the convex hexagon is equal to $(9/16) * \sqrt{3} - 3 * \sqrt{3}/32 = (15/32) * \sqrt{3}$.

The answer is (C).

Solution 2:

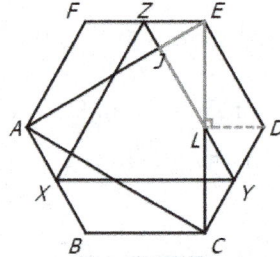

Similar to solution 1, subtract the area of three right triangles ZLJ from the equilateral triangle XYZ.

In the right triangle ELJ, $EL = \sqrt{3}/2$. Thus $CE = \sqrt{3}$ and $JE = \sqrt{3}/4$, so the area of equilateral triangle ACE is $(\sqrt{3}/4) * CE^2 = 3\sqrt{3}/4$.

In the right triangle ELJ, $\angle ELJ = 30°$ and $EJ = \sqrt{3}/4$. Thus the area of $\triangle ELJ$ is equal to $(1/2) * \sqrt{3}(EJ)^2 = 3\sqrt{3}/32$.

The area of the convex hexagon is equal to $(3/4) * \sqrt{3} - 3 * 3\sqrt{3}/32 = (15/32) * \sqrt{3}$.

The answer is (C).

Solution 3:

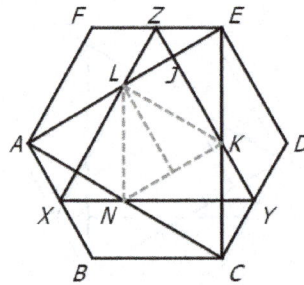

The targeted region can be split into five congruent right triangles as shown in the diagram and the right triangle is congruent to $Rt\triangle EJK$.

Similar to solution 2, we have $EK = \sqrt{3}/2$ and thus the area of $\triangle EJK$ is equal to $(\sqrt{3}/8) * (EK)^2 = (3/32) * \sqrt{3}$. Therefore, the area of the convex hexagon is equal to $(15/32) * \sqrt{3}$.

Answer is (C).

Problem 159

Sides AB and AC of equilateral triangle ABC are tangent to a circle at points B and C respectively. What fraction of the area of $\triangle ABC$ lies outside the circle? (2017 AMC10A)

(A) $4\sqrt{3}\pi/27 - 1/3$ (B) $\sqrt{3}/2 - \pi/8$ (C) $1/2$ (D) $\sqrt{3} - 2\sqrt{3}\pi/9$ (E) $4/3 - 4\sqrt{3}\pi/27$

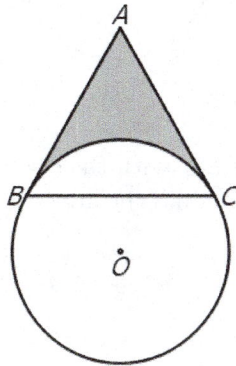

> **Tips:** *Connect the center of the circle to the tangent points to form right triangles.*

Solution 1:

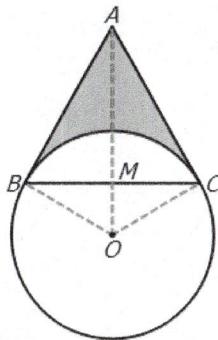

Connect OA, AB and OC. Then $\angle OBA = \angle OCA = 90°$ and $\angle BOC = 120°$. Let OC be 1. The area of section BOC is $\pi/3$. With $AC = \sqrt{3}$, the area of equilateral triangle ABC $S_{\triangle ABC} = 3\sqrt{3}/4$. The area of triangle OBC is $S_{\triangle OBC} = \sqrt{3}/4$.

Therefore, the area of the shaded region is $S_{\triangle ABC} + S_{\triangle OBC} - \pi/3 = \sqrt{3} - \pi/3$. The fraction of the area that lies outside the circle is $(\sqrt{3} - \pi/3)/(3\sqrt{3}/4) = 4/3 - 4\sqrt{3}\pi/27$.

The answer is (E).

Solution 2:

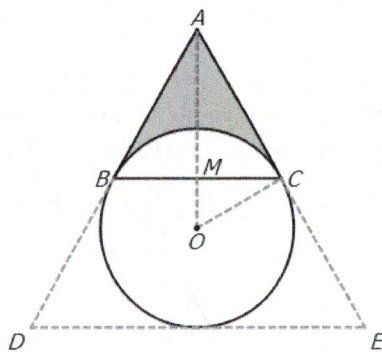

Construct an equilateral triangle ADC with the circle inscribed inside. Let OC be 1. The area of the inscribed circle is π. $AC = \sqrt{3}$ and the area of equilateral triangle ADE, $S_{\triangle ADE} = 3\sqrt{3}$.

The area of the shaded region is $(S_{\triangle ADE} - \pi)/3 = (3\sqrt{3} - \pi)/3$. Thus the fraction of the area that lies outside the circle is $(\sqrt{3} - \pi/3)/(3\sqrt{3}/4) = 4/3 - 4\sqrt{3}\pi/27$.

The answer is (E).

Problem 160

A square with side length x is inscribed in a right triangle with sides of length 3, 4, and 5 so that one vertex of the square coincides with the right-angle vertex of the triangle. A square with side length y is inscribed in another right triangle with sides of length 3, 4, and 5 so that one side of the square lies on the hypotenuse of the triangle. What is x/y ? (2017 AMC10A)

(A) 12/13   (B) 35/37   (C) 1   (D) 37/35   (E) 13/12

Tips: *Use similar triangles to calculate the edge-length of each square.*

Solution 1:

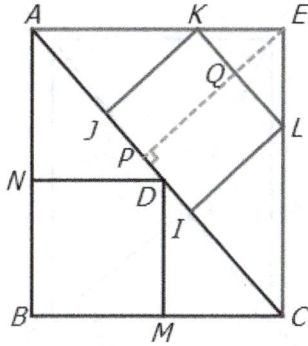

In the diagram above, $AB = EC = 4$, $AE = BC = 2$, and $AC = 5$.

Because $\triangle AND \sim \triangle ABC$, $ND/BC = AN/NB$. Thus $x/3 = (4-x)/4$. Solving the equation yields $x = 12/7$.

Because $\triangle EKL \sim \triangle EAC$, $KL/AC = EQ/EP$. Thus $y/5 = (12/5 - y)/(12/5)$. Solving the equation yields $y = 60/37$.

Therefore, the ratio $x/y = 37/35$.

The answer is (D).

Problem 161

Rectangle $ABCD$ has $AB = 3$ and $BC = 4$. Point E is the foot of the perpendicular from B to diagonal AC. What is the area of $\triangle AED$? (2017 AMC10B)

(A) 1   (B) 42/25   (C) 28/15   (D) 2   (E) 54/25

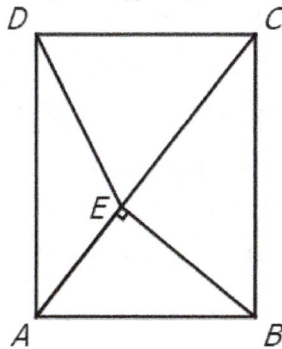

Tips: *Find the altitude to AD in the triangle ADE or use trigonometry.*

Solution 1:

241

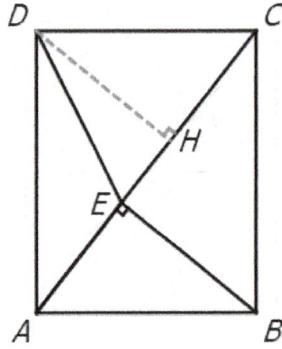

Let point H be on AC such that $DH \perp AC$. We have $\triangle ABC \sim \triangle AEB$ and $\triangle AEB \cong \triangle CHD$. In the right triangle ABC, $AB = 3$, $BC = 4$ and $AC = 5$, so $AE = AB^2/AC = 9/5$ and $DH = BE = 12/5$.

In the triangle ADE, $AE = 9/5$ and the length of altitude $DH = 12/5$. Therefore, the area of triangle ADE is equal to $54/25$.

The answer is (E).

Solution 2:

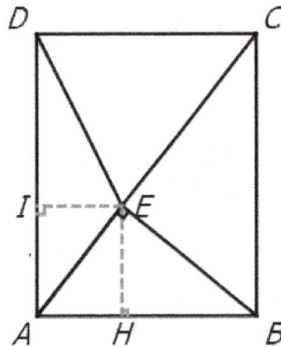

Let points H and I be on AB and AD such that $EH \perp AB$ and $EI \perp AD$.

Because BE is altitude in the right triangle ABC and EH is altitude in the right triangle ABE, $AE = AB^2/AC = 9/5$ and $AH = AE^2/AB = 27/25$.

Therefore, the area of triangle ADE is equal to $(1/2) * AD * AH = (1/2) \times 4 \times (27/25) = 54/25$

The answer is (E).

Solution 3:

242

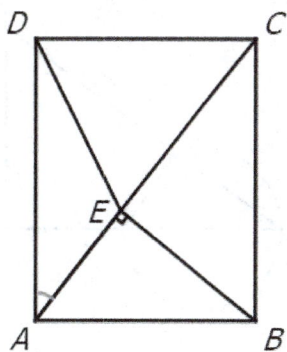

In the rectangle $AB = CD = 3$, $AD = BC = 4$, and $AC = 5$. Thus, $\sin \angle DAC = 3/5$.

Because BE is the altitude to AC in the right triangle, $AE = AB^2/AC = 9/5$. Therefore, the area of triangle ADE is equal to $(1/2)*AD*AE*\sin \angle DAC = (1/2) \times 4 \times (9/5) \times (3/5) = 54/25$.

The answer is (E).

Problem 162

The diameter AB of a circle of radius 2 is extended to a point D outside the circle so that $BD = 3$. Point E is chosen so that $ED = 5$ and line ED is perpendicular to line AD. Segment AE intersects the circle at a point C between A and E. What is the area of $\triangle ABC$? (2017 AMC10B)

(A) 120/37   (B) 140/39   (C) 145/39   (D) 140/37   (E) 120/31

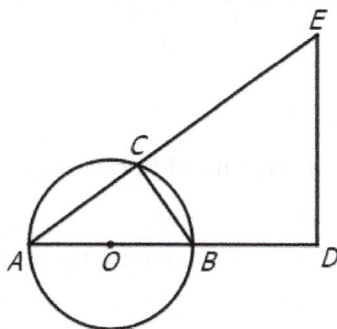

Tips: *Triangle ABC is similar to the triangle ADE.*

Solution 1:

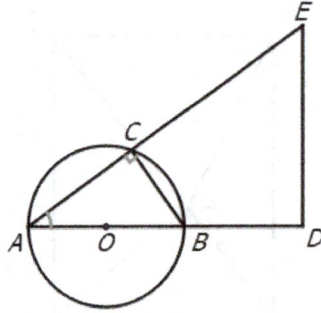

In the right triangle AED, $AD = 7$ and $ED = 5$. By the **Pythagorean Theorem**, $AE = \sqrt{74}$, so the area of triangle ADE is equal to $35/2$.

Connect BC and because AB is the diameter of the circle, $\angle ACB = 90°$. Thus $\triangle ABC \sim \triangle AED$ and then the area ratio between $\triangle ABC$ and $\triangle ADE$ is equal to $(AB/AE)^2 = (4/\sqrt{74})^2 = 8/37$.

Therefore, the area of $\triangle ABC$ is $(35/2) \times (8/37) = 140/37$.

The answer is (D).

Solution 2:

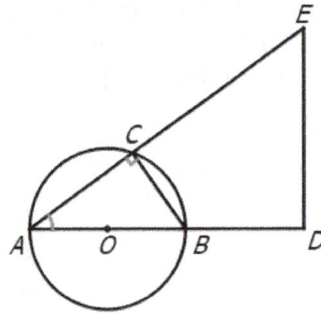

Similar to solution 1, the **Pythagorean Theorem** gives $AE = \sqrt{74}$. Thus, $\sin \angle EAD = 5/\sqrt{74}$ and $\cos \angle EAD = 7/\sqrt{74}$.

Connect BC. Because AB is the diameter of the circle, $\angle ACB = 90°$. Thus, $AC = AB \cos \angle EAD = 28/\sqrt{74}$.

The area of triangle ABC is equal to $(1/2) * AC * AB * \sin \angle EAD = (1/2) \times (28/\sqrt{74}) \times 4 \times (5/\sqrt{74}) = 140/37$.

The answer is (D).

Problem 163

Circles with centers P, Q and R, having radii 1, 2 and 3, respectively, lie on the same side of line AB and are tangent to AB at P', Q' and R', respectively, with Q' between P' and R'.

244

The circle with center Q is externally tangent to each of the other two circles. What is the area of $\triangle PQR$? (2016 AMC10A)

(A) 0   (B) $\sqrt{6}/3$ (C) 1   (D) $\sqrt{6} - \sqrt{2}$ (E) $\sqrt{6}/2$

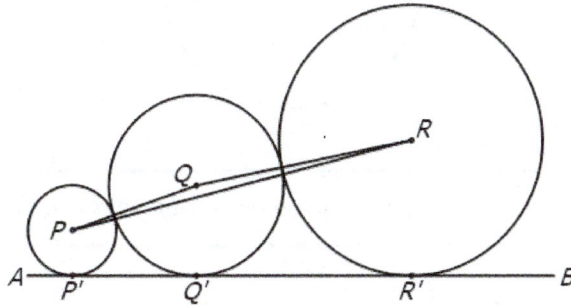

> **Tips:** *Connect the centers of the circles with each other and to the tangent points.*

Solution:

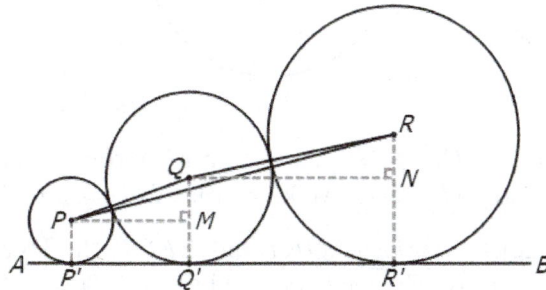

As shown in the diagram, there are three trapezoids, $PP'QQ'$, $QQ'RR'$, and $PP'RR'$ and the area of $\triangle PQR$ is equal to $S_{PP'QQ'} + S'_{QQ'RR} - S_{PP'RR'}$.

By the **Pythagorean Theorem**, in the right $\triangle PQM$, $PM = 2\sqrt{2}$ and in the right $\triangle QRN$, $QN = 2\sqrt{6}$.

Thus, $S_{PP'QQ'} = (PP' + QQ') * PM/2 = 3\sqrt{2}$, $S_{QQ'RR'} = (QQ' + RR') * QN/2 = 5\sqrt{6}$, and $S_{PP'RR'} = (PP' + RR') * (PM + QN)/2 = 4(\sqrt{2} + \sqrt{6})$.

Therefore, the area of $\triangle PQR$ is equal to $5\sqrt{6} + 3\sqrt{2} - 4(\sqrt{2} + \sqrt{6}) = \sqrt{6} - \sqrt{2}$.

The answer is (D).

Problem 164

A quadrilateral is inscribed in a circle of radius $200\sqrt{2}$. Three of the sides of this quadrilateral have length 200. What is the length of the fourth side? (2016 AMC10A)

(A) 200   (B) $200\sqrt{2}$ (C) $200\sqrt{3}$ (D) $300\sqrt{2}$ (E) 500

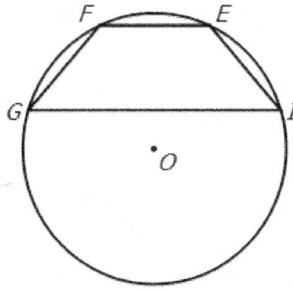

Tips: *Calculate the corresponding angle of each edge and use the Pythagorean Theorem.*

Solution 1:

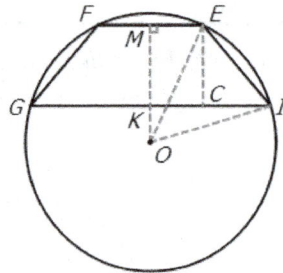

Because chords $GF = EI$, $FE // GI$. Let points M and C be on FE and GI, respectively, such that $OM \perp FE$ and $EC \perp GI$. We know that $ME = 100$. Let $GI = 200x$.

In the three right triangles $\triangle OME$, $\triangle OKI$ and $\triangle ECI$, the **Pythagorean Theorem** gives us $OM = 100\sqrt{7}$, $EC = 100\sqrt{2^2 - (x-1)^2}$, and $OK = 100\sqrt{2*2^2 - x^2}$.

We know that $OM = EC + OK$, so $\sqrt{4 - (x-1)^2} + \sqrt{8 - x^2} = \sqrt{7}$. Solving this equation yields $x = 5/2$. Therefore, $GI = 200x = 500$.

The answer is (E).

Solution 2:

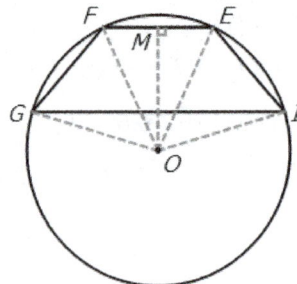

Connect the center of the circle to the vertices of the quadrilateral $GFEI$. Because $GF = FE = EI$, $\angle GOF = \angle FOE = \angle EOI$. let point M be on FE such that $OM \perp FE$. Thus, OM is the perpendicular bisector of FE and $\angle FOM = \angle MOE$. In the right triangle FOM, $\sin \angle FOM = (100/200) * \sqrt{2} = (1/2) * \sqrt{2}$.

246

By the **Triple Angle Formula** in trigonometry, $\sin \angle GOM = \sin(3\angle FOM) = 3\sin \angle FOM - 4(\sin \angle FOM)^3 = (3/2) * \sqrt{2} - (1/4) * \sqrt{2} = (5/4) * \sqrt{2}$. Therefore, $GI = 2OG \sin \angle GOM = (200\sqrt{2}) * (5/4) * \sqrt{2} = 500$.

The answer is (E).

Solution 3:

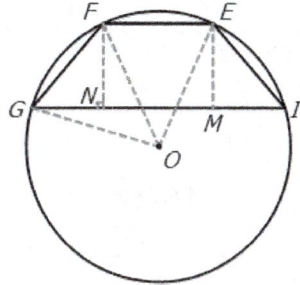

Similar to solution 1, $FE // GI$, so $\angle GFE + \angle FGI = 180°$. Thus, $\sin \angle GFE = \sin \angle FGI$.

In the isosceles triangle OFE, by the law of cosines, $\cos \angle OFE = FE^2/(2FE * OF) = (200/400) * \sqrt{2} = (1/2) * \sqrt{2}$. The Pythagorean identity gives us $\sin \angle OFE = (\sqrt{7}/2) * \sqrt{2}$.

Because $\triangle OGF \cong \triangle OFE$, $\angle OFG = \angle OFE$. Thus, $\sin \angle FGI = \sin \angle GFE = \sin(2\angle OFE) = (2 \times 1)/2\sqrt{2} \times (\sqrt{7}/2)\sqrt{2} = (1/4) * \sqrt{7}$. Similarly, $\cos \angle FGI = 3/4$.

Therefore, $GN = GF = 200 \times 3/4 = 150$ and $GI = 2GN + FE = 300 + 200 = 500$.

The answer is (E).

Problem 165

Rectangle $ABCD$ has $AB = 5$ and $BC = 4$. Point E lies on B so that $EB = 1$, point G lies on BC so that $CG = 1$. And point F lies on CD so that $DF = 2$. Segments AG and AC intersect EF at Q and P, respectively. What is the value of PQ/EF? (2016 AMC10B)

(A) $\sqrt{13}/16$ (B) $\sqrt{2}/13$ (C) $9/82$ (D) $10/91$ (E) $1/9$

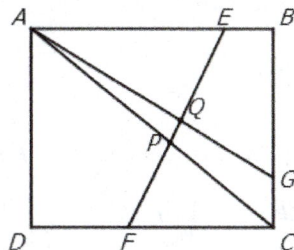

Tips: *Use pairs of similar triangles to calculate the length ratios on segment FE.*

Solution 1:

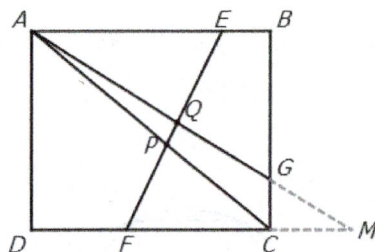

Extend AG and DC to meet at point M.

Because $\triangle AEP \sim \triangle CFP$, $EP/PF = AE/FC = 4/3$. Thus, $EP/EF = 4/7$.

Because $\triangle ABG \sim \triangle MCG$, $CM/AB = CG/GB = 1/3$. Thus, $CM = 5/3$ and $FM = 3 + 5/3 = 14/3$.

Because $\triangle AEP \sim \triangle MFQ$, $EQ/QF = AE/FM = 4/(14/3) = 6/7$. Thus, $EQ/EF = 6/13$.

Therefore, $PQ/EF = EP/EF - EQ/EF = 4/7 - 6/13 = 10/91$.

The answer is (D).

Solution 2:

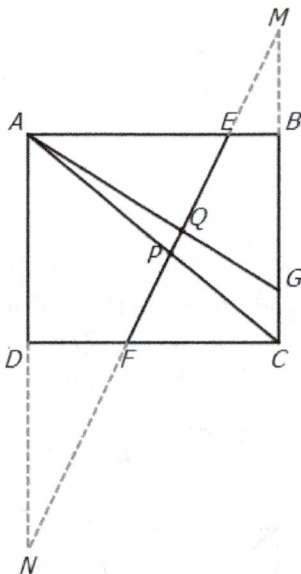

Extend FE and CB to meet at point M and extend EF and AD to meet at point N.

Because $\triangle AEP \sim \triangle CFP$, $EP/PF = AE/FC = 4/3$. Thus, $EP/EF = 4/7$.

Because $\triangle MEB \sim \triangle MFC$, $ME/MF = MB/MC = EB/FC = 1/3$. Thus, $MB = 2$ and $ME/EF = 1/2$.

Similarly, with $\triangle NDF \sim \triangle NAE$, $ND = 4$, $NF = EF$. Thus $MN = (5/2) * EF$.

Because $\Delta MQG \sim \Delta NQA$, $MQ/NQ = MG/AN = 5/8$. So $MQ = (5/13) * MN = (5/13) * (5/2) * EF = (25/26) * EF$ and $EQ = MQ - ME = 25/26 - 1/2 = 6/13$.

Therefore, $PQ/EF = EP/EF - EQ/EF = 4/7 - 6/13 = 10/91$.

The answer is (D).

Problem 166

What is the area of the region enclosed by the graph of the equation $x^2 + y^2 = |x| + |y|$? (2016 AMC10B)

(A) $\pi + \sqrt{2}$ (B) $\pi + 2$ (C) $\pi + 2\sqrt{2}$ (D) $2\pi + \sqrt{2}$ (E) $2\pi + 2\sqrt{2}$

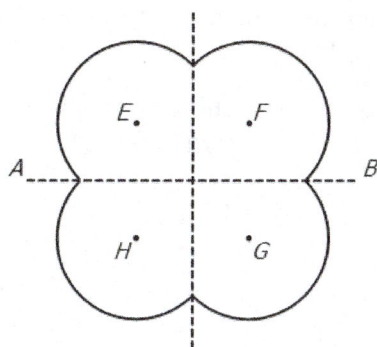

> **Tips:** 1. First draw the graph in quadrant I and then reflect over the x- and y-axis.

Solution 1:

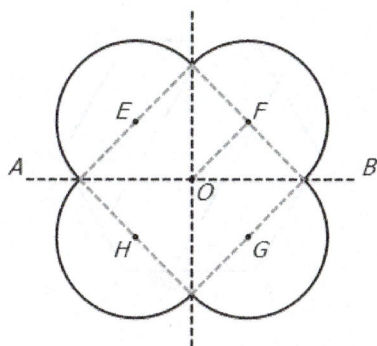

The original equation can be transformed into the equation $(|x| - 1/2)^2 + (|y| - 1/2)^2 = 1/2$. The center of the circles have the coordinates such that $|x| = 1/2$, $|y| = 1/2$ and radius of $\sqrt{2}/2$. The resulting curve is shown in the diagram and the edge-length of the square is $\sqrt{2}$.

Therefore, the total area of the area enclosed is $2\pi(\sqrt{2}/2)^2 + (\sqrt{2})^2 = \pi + 2$.

Answer is (B).

Solution 2:

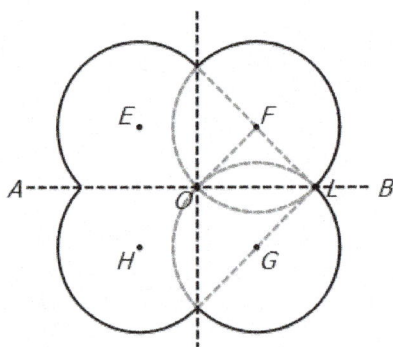

Similar to solution 1, the radius of the circle is $\sqrt{2}/2$ and the coordinates of the point F are $(1/2, 1/2)$. Thus, $\angle OFL = 90°$ and the area of the intersecting region between two circles is $2(\pi OF^2/4 - OF * LF/2) = \pi/4 - 1/2$.

The area of region enclosed is equal to the sum of four circles subtracted by 4 area of intersecting region, which is $4\pi/2 - 4(\pi/4 - 1/2) = \pi + 2$.

The answer is (B).

Problem 167

In regular hexagon $ABCDEF$, points W, X, Y, and Z are chosen on sides BC, CD, EF, and FA respectively, so lines AB, ZW, YX, and ED are parallel and equally spaced. What is the ratio of the area of hexagon $WCXYFZ$ to the area of hexagon $ABCDEF$? (2016 AMC10B)

(A) 1/3   (B) 10/27   (C) 11/27   (D) 4/9   (E) 13/27

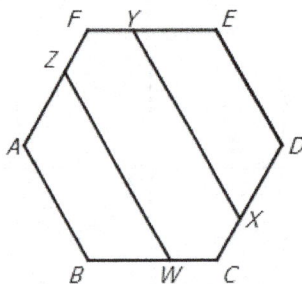

Tips: *Connect FC and calculate the areas of the resulting trapezoids.*

Solution 1:

250

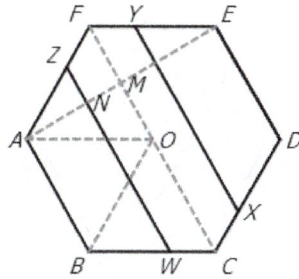

Connect AE and FC. Let AE intersect ZW at point N and intersect FC at point M. $ABCF$ and $ZWCF$ are trapezoids.

Because $AB//ZW//YX//ED$ and they are equally spaced, $MN = (1/2)*AN = (1/3)*AM$. Let the edge-length of the hexagon be x. We have $FC = 2x$ and $ZW = (5/3)x$.

The area of trapezoid $ABCF$ and $ZWCF$ are equal to $(x+2x)*AM/2$ and $(\frac{5}{3}x+2x)*NM/2$. Therefore, the area ratio of trapezoid $ZWCF$ to $ABCF$ is equal to $(11/6)/(3/2)*NM/AM = 11/27$.

The answer is (C).

Solution 2:

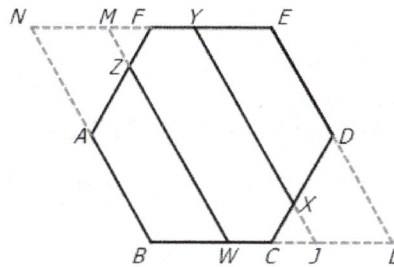

Extend BA and EF to meet at the point N and extend BC and ED to meet at point L.

Let the area of equilateral triangle CDL be x. Then, the area of $ABCDEF$ is $6x$, the area of $NBLE$ is 8x, and the area of $MWJY$ is $\frac{8}{3}x$.

Because the parallel lines AB, ZW, YX and ED are equally spaced, $CJ = JL/2 = CL/3$. Thus, the area of equilateral triangle XCJ is $\frac{1}{9}x$.

The total area of $ZWCXYF$ is $\frac{8}{3}x - 2*\frac{1}{9}x = \frac{22}{9}x$. Therefore, the targeted area ratio is $(22/9)/6 = 11/27$.

The answer is (C).

Solution 3:

251

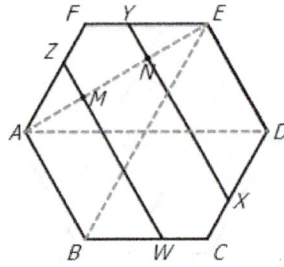

Let the edge-length of the hexagon be 1. The total area of $ABCDEF$ is $6\sqrt{3}/4 = 3\sqrt{3}/2$. And $AE = \sqrt{3}$. The equal spacing gives us $EN = NM = \sqrt{3}/3$.

In right triangle ENY, $\angle NEY = 30°$, so $YN = 1/3$ and $YX = 1/3 + 1 + 1/3 = 5/3$. Thus, the area of rectangle $ZWXY$ is equal to $5\sqrt{3}/9$. In the isosceles triangle FZY, $\angle NEY = 120°$ and $ZY = MN = \sqrt{3}/3$. Thus, the area of triangle FZY is equal to $\sqrt{3}/36$ and the total area of $ZWCXYF$ is $5\sqrt{3}/9 + 2 \times \sqrt{3}/36 = 11\sqrt{3}/18$.

The targeted area ratio is $(11\sqrt{3}/18)/(3\sqrt{3}/2) = 11/27$.

The answer is (C).

Problem 168

Tetrahedron $ABCD$ has $AB = 5$, $AC = 3$, $BC = 4$, $BD = 4$, $AD = 3$, and $CD = 12/5\sqrt{2}$. What is the volume of the tetrahedron? (2015 AMC10A)

(A) $3\sqrt{2}$ (B) $2\sqrt{5}$ (C) $24/5$ (D) $3\sqrt{2}$ (E) $24/5\sqrt{2}$

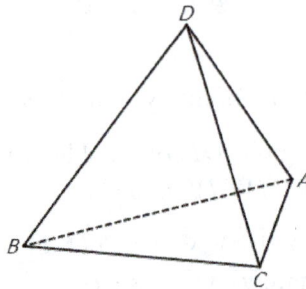

Tips: *Connect midpoint of CD to A and B to form a triangle as the base.*

Solution:

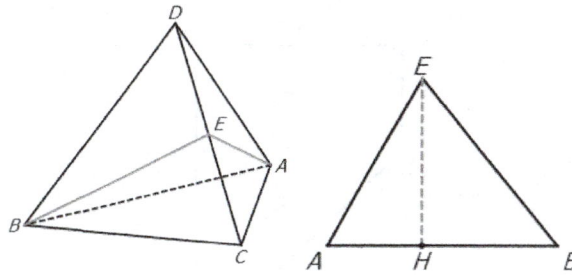

Let E be the midpoint of CD and connect BE and AE.

Because $BC = BD$, $AD = AC$ and $DE = EC$, $BE \perp CD$ and $AE \perp CD$. Thus, the volume of the tetrahedron $ABCD$ is equal to $(1/3) * CD * S_{\triangle ABE}$.

By the **Pythagorean Theorem**, in right triangle BCE and ACE, $BE^2 = BC^2 - CE^2 = 4^2 - 72/25$ and $AE^2 = AC^2 - CE^2 = 3^2 - 72/25$.

In the triangle ABE, EH is the altitude of side AB. Let $x = AH$. Then, $BH = 5 - x$. By the **Pythagorean Theorem**, $AE^2 - x^2 = BE^2 - (5 - x)^2$. Substituting AE and BE yields $9 - x^2 = 16 - (5 - x)^2$. Solving the equation gives $x = 9/5$. Thus $EH = 6\sqrt{2}/5$ and the area of triangle ABE is $S_{\triangle ABE} = 3\sqrt{2}$.

Therefore, the volume of tetrahedron is $(1/3) * (12\sqrt{2}/5) * 3\sqrt{2} = 24/5$.

Problem 169

For some positive integers p, there is a quadrilateral $ABCD$ with positive integer side lengths, perimeter p, right angles at B and C, $AB = 2$, and $CD = AD$. How many different values of $p < 2015$ are possible? (2015 AMC10A)

(A) 330   (B) 340   (C) 350   (D) 360   (E) 370

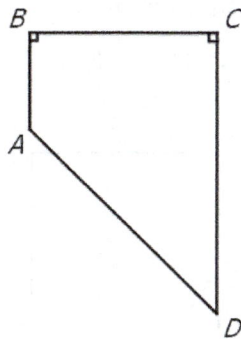

Tips: *Establish constraints for segment CD through the Pythagorean Theorem.*

Solution:

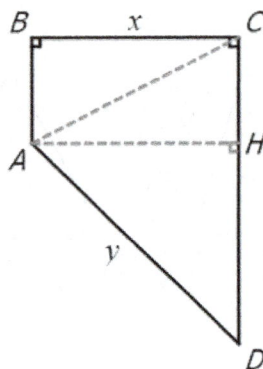

Let $x = BC$ and $y = AD = CD$. Thus, $p = 2 + x + 2y < 2015$.

In the $\triangle ACD$, AH is the altitude of side CD. By the **Pythagorean Theorem**, $AD^2 = AH^2 + DH^2$ which is $y^2 = x^2 + (y-2)^2$. Simplifying the equation yields $y = 1 + x^2/4 \geq 2$. Substituting it into the inequality gives $p = 2 + x + 2 + x^2/2 < 2015$. We have two inequalities:

$1 + x^2/4 \geq 2$ (a)

$p = 2 + x + 2 + x^2/2 < 2015$ (b)

The inequality (a) yields $x \geq 2$ and inequality (b) yields $x < 62.4$. We know that x must be an even number in order to have y be an integer.

The number of even numbers between 2 and 62.4 is 31.

The answer is (B).

Problem 170

Let S be a square of side length 1. Two points are chosen at random on the sides of S. The probability that the straight-line distance between the points is at least $1/2$ is $(a - b\pi)/c$, where a, b, and c are positive integers with $gcd(a, b, c) = 1$. What is $a + b + c$? (2015 AMC10A)

(A) 59   (B) 60   (C) 61   (D) 62   (E) 63

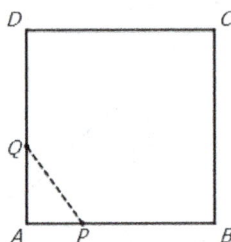

Tips: *Use the method of casework: both points on the same edge, on opposite edges, or on adjacent edges.*

254

Solution:

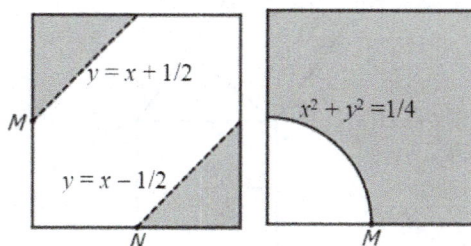

There are four scenarios: the two points are on the same edge (Case I), the two points are on adjacent edges (Case II and III), or the two points are on opposite edges (Case IV). The probability of each case is $1/4$.

Case I: the two points are on the same edge. The edge is represented by $[0,1]$. Let x and y be the coordinates along $[0,1]$ for the two points. Then, the constraint is $|x - y| \geq 1/2$, which is equivalent to $y \geq x + 1/2$ or $y \geq x - 1/2$. The probability is equal to the shaded area in the diagram (left), which is $1/4$.

Case II and III: the two points are on adjacent edges. Let the coordinates of the two points be $(x, 0)$ and $(0, y)$. The constraint is $\sqrt{x^2 + y^2} \geq 1/2$, which is can be written as $x^2 + y^2 \geq (1/2)^2$. The probability is denoted by the shaded area in the diagram (right), which is $1 - \pi/16$.

Case IV: two points are on opposite edges. It is easy to see that the distance between the two points is greater than one, so the probability that the distance is greater than $1/2$ is 1.

Therefore, the probability that the straight-line distance between the points is at least $1/2$ is $(1/4) \times (1/4 + 2(1 - \pi/16) + 1) = (26 - \pi)/32$.

$a + b + c = 26 + 1 + 32 = 59$.

The answer is (A).

Problem 171

In $\triangle ABC$, $\angle C = 90°$ and $AB = 12$. Squares $ABXY$ and $ACWZ$ are constructed outside of the triangle. The points X, Y, Z, and W lie on a circle. What is the perimeter of the triangle? (2015 AMC10B)

(A) $12 + 9\sqrt{3}$ (B) $18 + 6\sqrt{3}$ (C) $12 + 12\sqrt{2}$.   (D) 30 (E) 32

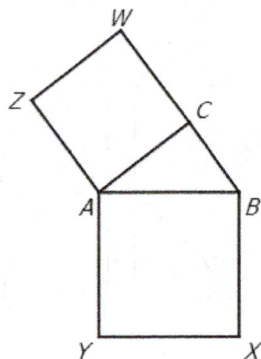

Tips: *First prove that triangles AZY and BWX are similar.*

Solution:

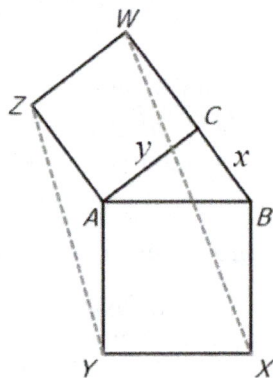

Connect ZY and WX. Let $x = BC$ and $y = AC$.

Because X, Y, Z and W lie on a circle, $\angle ZWX + \angle XYZ = 180°$ and $\angle WZY + \angle WXY = 180°$. In square $ABXY$ and $ACWZ$, $\angle ZWC = \angle WZA = \angle BXY = \angle XYZ = 90°$. Thus, $\angle YZA = \angle WXB$ and $\angle ZYA = \angle XWB$, so $\triangle ZAY \sim \triangle XBW$.

Because $\triangle ZAY \sim \triangle XBW$, $ZA/BX = YA/WB$ and $y/12 = 12/(x+y)$. By the **Pythagorean Theorem**, we have $x^2 + y^2 = 12^2$. Solving the equation set yields $x = y = 6\sqrt{2}$. Thus, the perimeter of the triangle is $12 + 12\sqrt{2}$.

The answer is (C).

Problem 172

In the figure shown below, $ABCDE$ is a regular pentagon and $AG = 1$. What is $FG + JH + CD$? (2015 AMC10B)

(A) 3   (B) $12 - 4\sqrt{5}$ (C) $(5 + 2\sqrt{5})/3$ (D) $1 + \sqrt{5}$ (E) $(11 + 11\sqrt{5})/10$

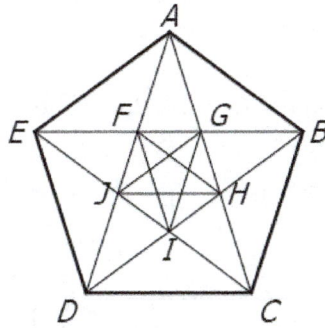

Tips: *Calculate the length of EG through similar triangles.*

Solution 1:

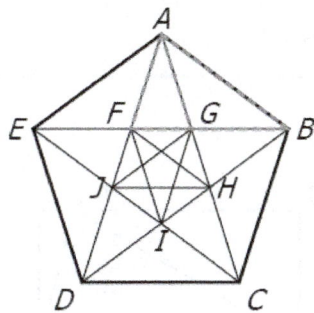

In the regular pentagon, $\angle FAG = \angle GAB = \angle GBA = 36°$ and $AB = BF$, $AF = AG = JH = 1$.

Let $x = FG$. Thus, $CD = AB = BF = 1 + x$ and $FG + JH + CD = 2 + 2x$.

The triangle AFG is a golden triangle, which means $x = (\sqrt{5} - 1)/2$. Therefore, $FG + JH + CD = 2 + 2x = 1 + \sqrt{5}$.

The answer is (D).

Solution 2:

In the regular pentagon, $\angle FAG = \angle GAB = \angle GBA = 36°$ and $AB = BF$, $AF = AG = 1$. Thus, AG is angular bisector of $\angle FAB$ and $\triangle AFG \sim \triangle BFA$.

Let $x = FG$. Thus, $AB = BF = 1 + x$. The relationship $AB/AF = AF/FG$ yields $(1 + x)/1 = 1/x$, so $x^2 + x - 1 = 0$. Solving the equation gives us $x = (\sqrt{5} - 1)/2$.

Because $\triangle AFG \sim \triangle AJH \sim \triangle ADC$, $JH = AH/AG \cdot FG = (1+x)x$ and $DC = AC/AG \cdot FG = (2 + x)x$. Thus, $FG + JH + CD = 2x^2 + 4x = 1 + \sqrt{5}$.

The answer is (D).

Problem 173

In rectangle $ABCD$, $AB = 20$ and $BC = 10$. Let E be a point on CD such that $\angle CBE = 15°$. What is AE? (2014 AMC10A)

(A) $20\sqrt{3}/3$ (B) $10\sqrt{3}$ (C) 18   (D) $11\sqrt{3}$ (E) 20

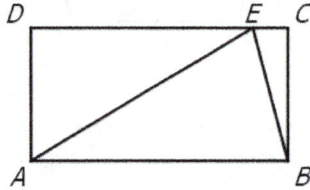

Tips: *Calculate the length of CE through constructing a right triangle with one 30 degree angle and CD as a leg.*

Solution 1:

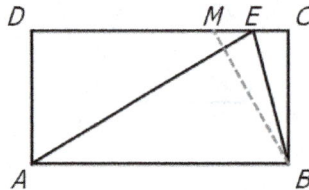

Let point M be on CD such that $\angle CBM = 30°$. Thus, EB is the angular bisector of $\angle CBM$. In the $30-60-90$ right triangle CBM, $MB = 2/\sqrt{3} * BC = 20/\sqrt{3}$ and $MC = \sqrt{3}/3 * BC = 10\sqrt{3}/3$.

By the **Angular Bisector Theorem**, $CE = BC/(BC + MB) * MC = \sqrt{3}/(2 + \sqrt{3}) * 10\sqrt{3}/3 = 20 - 10\sqrt{3}$. $DE = CD - CE = 10\sqrt{3}$.

The answer is (E).

Solution 2:

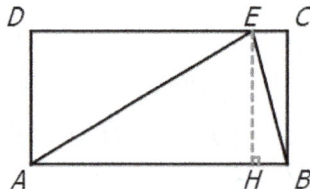

In this solution, we will use trigonometry to find CE.

We know that $CE = BC \tan 15° = 10 \tan 15°$.

258

$\tan 15° = \tan (45° - 30°) = (\tan 45° - \tan 30°)/(1 + \tan 45° \tan 30°) = (1 - \sqrt{3}/3)/(1 + \sqrt{3}/3) = 2 - \sqrt{3}$.

Therefore, $CE = 20 - 10\sqrt{3}$. The rest of the steps will be the same as in solution 1.

The answer is (E).

Problem 174

Trapezoid $ABCD$ has parallel sides AB of length 33 and CD of length 21. The other two sides are of lengths 10 and 14. The angles at A and B are acute. What is the length of the shorter diagonal of $ABCD$? (2014 AMC10B)

(A) $10\sqrt{6}$ (B) 25   (C) $8\sqrt{10}$ (D) $18\sqrt{2}$ (E) 26

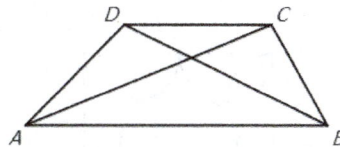

Tips: *Construct a parallelogram to transfer all the length information into one triangle.*

Solution 1:

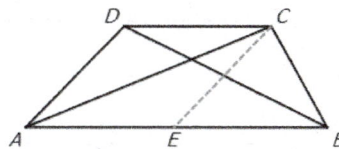

Let point E be on AB such that $CE//AD$. Thus, $ADCE$ is a parallelogram and $CE = AD = 14$ and $BE = AB - CD = 12$.

By the law of cosines, $\cos \angle B = (10^2 + 12^2 - 14^2)/(2 * 12 * 10) = 2/5$.

Thus, $\cos C = -\cos B = -2.5$.

Applying the law of cosines in triangle ABC and BCD yields $AC^2 = AB^2 + BC^2 - 2AB * BC * cos(B) = 725$ and $BD^2 = CD^2 + BC^2 - 2CD * BC * \cos C = 625$. Thus, the shorter diagonal is $BD = 25$.

The answer is (B).

Solution 2:

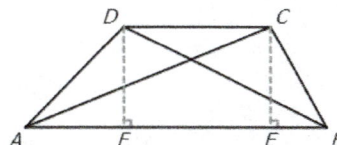

259

Let points E and F be on segment AB such that $CE \perp AB$ and $DF \perp AB$. Let $x = AF$ and $y = BE$. Thus, $x + y = 33 - 21 = 12$.

By the **Pythagorean Theorem**, $AD^2 - AF^2 = BC^2 - BE^2 = DF^2$ which gives $14^2 - x^2 = 10^2 - y^2$. Solving the equation set gives $x = 10$ and $y = 2$ and $DF = 4\sqrt{6}$. The shorter diagonal is $BD = \sqrt{96 + 23^2} = 25$.

The answer is (B).

Problem 175

Eight semicircles line the inside of a square with side length 2 as shown. What is the radius of the circle tangent to all of these semicircles? (2014 AMC10B)

(A) $(1 + \sqrt{4})/4$ (B) $(\sqrt{5} - 1)/2$ (C) $\sqrt{3}$ (D) $2\sqrt{3}/5$ (E) $\sqrt{5}/3$

> **Tips:** *Construct a right triangle by connecting the centers of the semicircles.*

Solution 1:

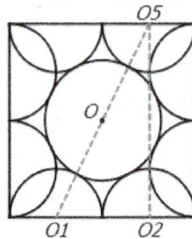

Connect the centers of semicircles O_1O_5 and O_5O_2 as shown in the diagram. We have a right triangle $O_1O_2O_5$ with O_1O_5 passing through O.

It is easy to find that $O_1O_2 = 1$ and $O_2O_5 = 2$. Thus, $O1O5 = \sqrt{5}$ and the radius of the center circle is $(\sqrt{5} - 1)/2$.

The answer is (B).

Solution 2:

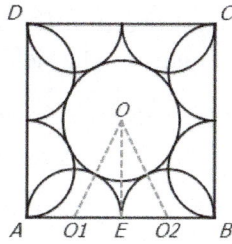

Connect OO_1 and OE as shown in the diagram. $O_1E = 1/2$ and $OE = 1$. Thus, $OO_1 = \sqrt{5}/2$. Therefore, the radius of center circle is $\sqrt{5}/2 - 1/2$.

The answer is (B).

Problem 176

A sphere is inscribed in a truncated right circular cone as shown. The volume of the truncated cone is twice that of the sphere. What is the ratio of the radius of the bottom base of the truncated cone to the radius of the top base of the truncated cone? (2014 AMC10B)

(A) 3/2   (B) $(1+\sqrt{5})/2$ (C) $\sqrt{3}$ (D) 2   (E) $(3+\sqrt{5})/2$

> **Tips:** 1. Cut through the axis of the cone to form a trapezoid with an inscribed circle.
> 2. Establish two equations involving radius of top base, bottom base and inscribed circle.

Solution:

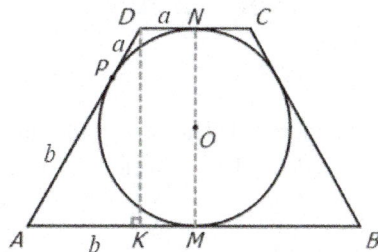

Let the radius of the top base, the bottom base, and the inscribed sphere be a, b and r, respectively. $NM = DK = 2r$. The information about the ratio of the volumes yields $2 * (4/3) * \pi r^3 = (1/3) * \pi * (2r) * (a^2 + ab + b^2)$. Therefore, $4r^2 = a^2 + ab + b^2$ (1)

261

In the right triangle ADK, the **Pythagorean Theorem** gives us $(a+b)^2 = (b-a)^2 + (2r)^2$. Thus, $4r^2 = 4ab$. Substituting into equation (1) yields $a^2 - 3ab + b^2 = 0$.

Solving the equation gives $a/b = (3+\sqrt{5})/2$

The answer is (E).

Problem 177

A unit square is rotated $45°$ about its center. What is the area of the region swept out by the interior of the square? (2013 AMC10A)

(A) $1 - \sqrt{2}/2 + \pi/4$ (B) $1/2 + \pi/4$ (C) $2 - \sqrt{2} + \pi/4$ (D) $\sqrt{2}/2 + \pi/4$ (E) $1 + \sqrt{2}/4 + \pi/8$

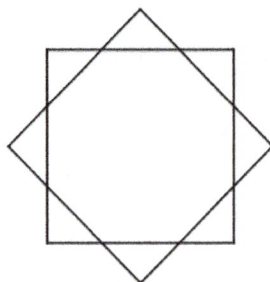

> **Tips:** *Draw the diagram carefully and identify the intersecting points of the edges.*

Solution 1:

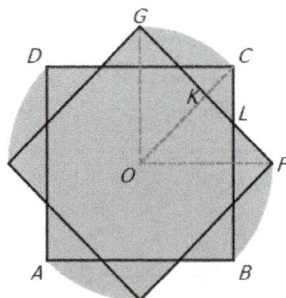

The area swept by the square is shown in the diagram. The region has 4 sectors congruent to GOC, each with an area equal to $1/8$ of the circle and four pieces congruent to $OCLF$.

The radius of the circle is $\sqrt{2}/2$, $KF = 1/2$ and $LF = \sqrt{2}KL$. Thus $KL = (\sqrt{2}-1)/2$. So the area of triangle OKF and KCL is $1/8$ and $(3-2\sqrt{2})/8$. Thus, the area of $OCLF$ is $1/2 - \sqrt{2}/4$.

The swept out area is $\pi/4 + 2 - \sqrt{2}$.

The answer is (C).

Solution 2:

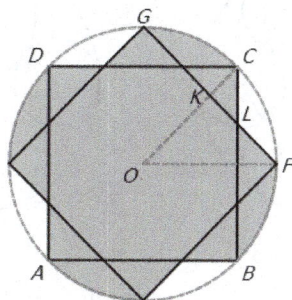

The swept area is the difference between the area of the circle and the four small unshaded pieces (CLF). The area of the circle is $\pi/2$.

The area of the sector COF is $\pi/8$ and $KF = 1/2 = (1 + \sqrt{2})KL$. Thus, $KL = \sqrt{2} - 1$. The areas of KLC and OKF are $(3 - 2\sqrt{2})/2$ and $1/8$, respectively. Therefore, the area of unshaded area LCF is equal to $\pi/16 - 1/8 - (3 - 2\sqrt{2})/2$.

The total area of shaded area is equal to $\pi/2 - 4(\pi/16 - 1/8 - (3 - 2\sqrt{2})/2) = \pi/4 + 2 - \sqrt{2}$.

The answer is (C).

Problem 178

Six spheres of radius 1 are positioned so that their centers are at the vertices of a regular hexagon of side length 2. The six spheres are internally tangent to a larger sphere whose center is the center of the hexagon. An eighth sphere is externally tangent to the six smaller spheres and internally tangent to the larger sphere. What is the radius of this eighth sphere? (2013 AMC10A)

(A) $\sqrt{2}$ (B) $3/2$ (C) $5/3$ (D) $\sqrt{3}$ (E) 2

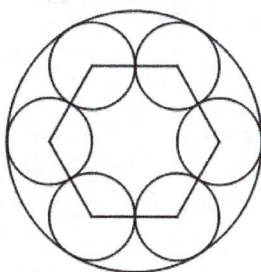

> **Tips:** *Create a cutting plan through two small diagonal spheres and two bigger spheres to form right triangles.*

Solution:

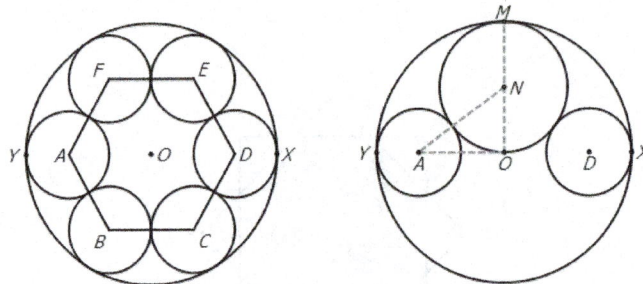

The top view of the hexagon and six spheres with externally tangent sphere is shown as the diagram (left). $OA = 2$ and $OY = 3$.

Cutting through line XY results in the side view as shown on the right. $OM = OY = 3$. Let the radius $MN = r$. In the right triangle, we have $AN^2 = ON^2 + OA^2$, which is $(1+r)^2 = 2^2 + (3-r)^2$. Solving the equation yields $r = 3/2$.

The answer is (B).

Problem 179

In $\triangle ABC$, $AB = 86$, and $AC = 97$. A circle with center A and radius AB intersects BC at points B and X. Moreover BX and CX have integer lengths. What is BC? (2013 AMC10A)

(A) 11   (B) 28   (C) 33   (D) 61   (E) 72

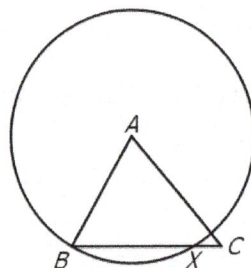

Tips: *Establish a constraint by the Power of a Point Theorem.*

Solution:

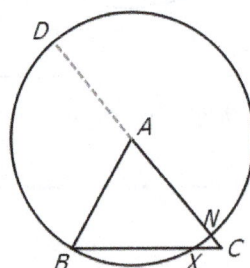

Extend CA to meet circle at point D. Then $CD = 86 + 97 = 183$ and $CN = 97 - 86 = 11$.

Let $x = CB$ and $y = CX$. We have the constraints $x > y$ and $x < 183$. By **the Power of a Point Theorem**, $CN * CD = xy$. Thus, $xy = 11 \times 3 \times 61$, where x and y are integers such that $x > y$ and $x < 183$. Therefore, $x = 61$ and $y = 33$.

The answer is (D).

Problem 180

In triangle ABC, medians AD and CE intersect at P, $PE = 1.5$, $PD = 2$, and $DE = 2.5$. What is the area of $AEDC$? (2013 AMC10B)

(A) 13   (B) 13.5   (C) 14   (D) 14.5   (E) 15

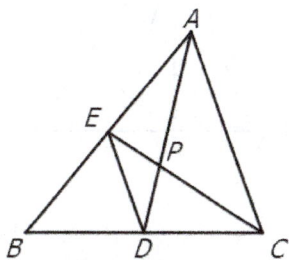

> **Tips:** *Find the type of triangle of APC.*

Solution:

Because AD and CE are medians of triangle ABC, DE is midline and P is the centroid of triangle ABC. Thus, $AC = 2DE = 5$, $AP = 2PD = 4$, and $CP = 2PE = 3$. By the converse of the **Pythagorean Theorem**, $AP \perp CP$.

Therefore, the area of $AEDC$ is $(1/2) \times 6 \times 4.5 = 13.5$.

The answer is (B).

Problem 181

In triangle ABC, $AB = 13$, $BC = 14$, and $CA = 15$. Distinct points D, E, and F lie on segments BC, CA, and DE, respectively, such that $AD \perp BC$, $DE \perp AC$, and $AF \perp BF$. The length of segment DF can be written as m/n, where m and n are relatively prime positive integers. What is $m + n$? (2013 AMC10B)

(A) 18   (B) 21   (C) 24   (D) 27   (E) 30

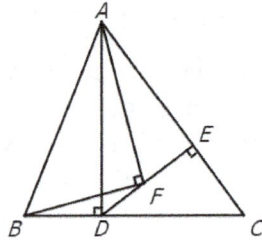

Tips: *Prove that* $\triangle ABD \sim \triangle AEF$ *or* $\triangle ABF \sim \triangle ADE$.

Solution :

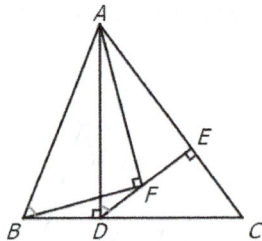

Let $x = BD$ and $y = CD$. Thus, $x + y = 14$. By the **Pythagorean Theorem**, $AC^2 - AB^2 = CD^2 - BD^2$, which gives the equation $15^2 - 13^2 = 14(y - x) = 56$. Solving the equations gives us $x = 5$, $y = 9$, and $AD = 12$.

Because in the right triangle ACD, $DE \perp AC$, $AE = AD/AC = 48/5$, $CE = AC - AE = 27/5$ and $DE = 36/5$.

Because $AF \perp BF$ and $AD \perp BD$, quadrilateral $ABDF$ is concyclic and $\angle ABF = \angle ADE$. Then $\triangle ABF \sim \triangle ADE$, which yields $AF/AE = AB/AD$. So $AF = AE * AB/AD = (48/5) * 13/12 = 52/5$.

By the **Pythagorean Theorem**, in triangle AEF, $EF = 4$. Therefore, $DF = 36/5 - 4 = 16/5$, so $m + n = 21$.

The answer is (B).

Solution 2:

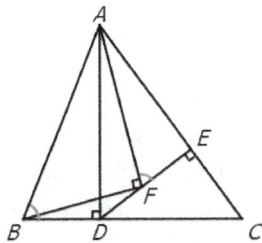

Similar as in solution 1, the **Pythagorean Theorem** gives $AC^2 - AB^2 = CD^2 - BD^2$, which results in the equation $15^2 - 13^2 = 14(y - x) = 56$. Solving gives us $x = 5$, $y = 9$, and $AD = 12$.

266

Because in the right triangle ACD, $DE \perp AC$, $AE = AD/AC = 48/5$, $CE = AC - AE = 27/5$ and $DE = 36/5$.

Because $AF \perp BF$ and $AD \perp BD$, points $ABDF$ is concyclic and $\angle ABD = \angle AFE$. Thus $\triangle ABD \sim \triangle AFE$ which yields $FE/AE = BD/AD$. So $FE = BD * AE/AD = ((5 * 36)/5)/9 = 4$. Therefore, $DF = 36/5 - 4 = 16/5 = m/n$. $m + n = 21$.

The answer is (B).

Solution 3:

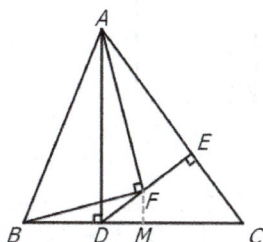

Similar t in solution 1, the **Pythagorean Theorem** gives $AC^2 - AB^2 = CD^2 - BD^2$, which results in the equation $15^2 - 13^2 = 14(y - x) = 56$. Solving gives us $x = 5$, $y = 9$, and $AD = 12$.

Because in the right triangle ACD, $DE \perp AC$, $AE = AD/AC = 48/5$, $CE = AC - AE = 27/5$ and $DE = 36/5$.

M is on CD such that $FM \perp CD$. Let $z = DF$, thus $FM = (3/5) * z$, $DM = (4/5) * z$ and $EF = 36/5 - z$. Applying the **Pythagorean Theorem** in right triangle ABF, we get $132 = BF^2 + AF^2 = (5 + (4/5) * z)^2 + ((3/5) * z)^2 + (48/5)^2 + (36/5 - z)^2$. Simplifying and solving the equation gives $z = 16/5 = m/n$. So $m + n = 21$.

The answer is (B).

Problem 182

Three unit squares and two line segments connecting two pairs of vertices are shown. What is the area of $\triangle ABC$? (2012 AMC10A)

(A) 1/6   (B) 1/5   (C) 2/9   (D) 1/3   (E) $\sqrt{2}/4$

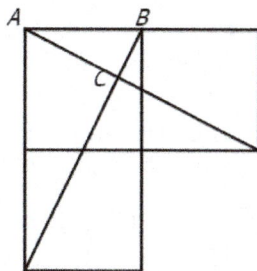

Tips: *Find similar triangle pairs in the diagram.*

Solution 1:

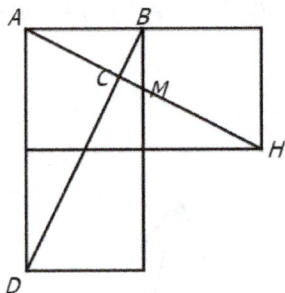

It is easy to see that $\triangle ACD \sim \triangle MCB$. We get the equation $AD/BM = CA/CM$, which yields $2/0.5 = AC/CM = 4$. Thus, $AC/AM = 4/5$. The area of triangle $ABM = 1/4$. The area of triangle ABC is equal to $(4/5)*(1/4) = 1/5$.

The answer is (B).

Solution 2:

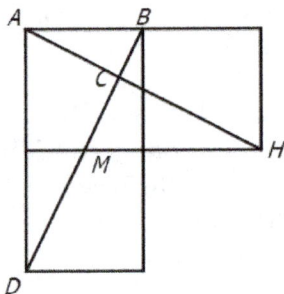

It is straightforward to find $\triangle ACB \sim \triangle MCB$. $BC/CM = AB/MN$ yields $BC/CM = 1/1.5 = 2/3$. So $BC/BM = 2/5$ and $BC/BD = 1/5$.

The area of triangle ABD is 1. Therefore, the area of triangle ABC is equal to $1*(1/5) = 1/5$.

The answer is (B).

Problem 183

Let points $A = (0,0,0)$, $B = (1,0,0)$, $C = (0,2,0)$, and $D = (0,0,3)$. Points E, F, G, and H are midpoints of line segments BD, AB, AC, and DC, respectively. What is the area of $EFGH$? (2012 AMC10A)

(A) $\sqrt{2}$ (B) $2\sqrt{5}/3$ (C) $3\sqrt{5}/4$ (D) $\sqrt{3}$ (E) $2\sqrt{7}/3$

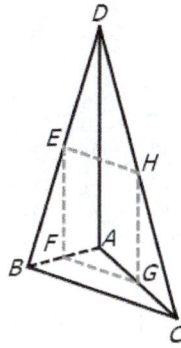

Tips: *Recognize that EFGH is a rectangle.*

Solution:

From the coordinates of points A, B, C and D, we have $AB \perp AC$ and $AD \perp \triangle ABC$.

By the **Pythagorean Theorem** in triangle AFG, $FG = \sqrt{5}/2$. Because FE is a midline in triangle ABD, $EF//AD$. Similarly, $HG//AD$. Therefore, $EF \perp \triangle ABC$, $HG \perp \triangle ABC$, and $EF = HG = AD/2 = 1.5$. Because $EF \perp FG$ and $HG \perp FG$, $EFGH$ is a rectangle and its area is $FG * HG = 3\sqrt{5}/4$.

The answer is (C).

Problem 184

Three circles with radius 2 are mutually tangent. What is the total area of the circles and the region bounded by them, as shown in the figure? (2012 AMC10B)

(A) $10\pi + 4\sqrt{2}$ (B) $13\pi - \sqrt{3}$ (C) $12\pi + \sqrt{3}$ (D) $10\pi + 9$ (E) 13π

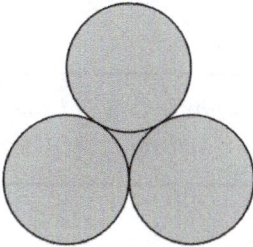

Tips: *Connect the center of the circles to form an equilateral triangle.*

Solution:

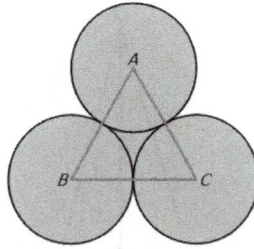

Connect the centers of three circles to form an equilateral $\triangle ABC$ with side-length of 4. The area of the shaded region is equal to the sum of area of the three pies with a 300 degree central angle and equilateral $\triangle ABC$. Therefore, the area of the shaded region is equal to $(5/6) * 4\pi \times 3 + \sqrt{3}/4 \times 4^2 = 10\pi + 4\sqrt{3}$.

The answer is (A).

Problem 185

Jesse cuts a circular paper disk of radius 12 along two radii to form two sectors, the smaller having a central angle of 120 degrees. He makes two circular cones, using each sector to form the lateral surface of a cone. What is the ratio of the volume of the smaller cone to that of the larger? (2012 AMC10B)

(A) 1/8   (B) 1/4   (C) $\sqrt{10}/10$ (D) $\sqrt{5}/6$ (E) $\sqrt{5}/5$

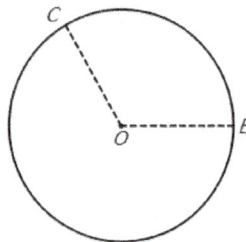

> **Tips:** 1. The perimeter of base circle is equal to the length of the arc cut from original disk. 2. The hypotenuse of the resulting right triangle is the radius of original disk.

Solution:

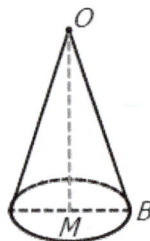

270

The cone constructed is shown in the diagram. The perimeter of the base circle is equal to the length of arc cut from the disk. Let the radius of the base circle be a and b for the cones resulting from 120° and 240° pie.

We have the equations $2\pi * a = (1/3) * 2\pi r$ and $2\pi * b = (2/3) * 2\pi r$. Solving the equations gives us $a = r/3 = 4$ and $b = 2r/3 = 8$. The altitude of the resulting cones is $8\sqrt{2}$ and $4\sqrt{5}$, respectively. Therefore, the volume ratio is $(a^2 * 8\sqrt{2})/(b^2 * 4\sqrt{5}) = \sqrt{10}/10$.

The answer is (C).

Problem 186

In rectangle $ABCD$, $AB = 6$, $AD = 30$, and G is the midpoint of AD. Segment AB is extended 2 units beyond B to point E, and F is the intersection of ED and BC. What is the area of $BFDG$? (2012 AMC10B)

(A) 133/2   (B) 67   (C) 135/2   (D) 68   (E) 137/2

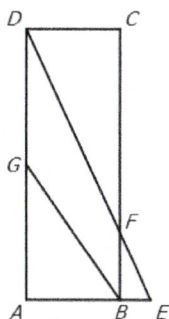

Tips: *Use similar triangles to find the length of BF and calculate the area f the trapezoid.*

Solution 1:

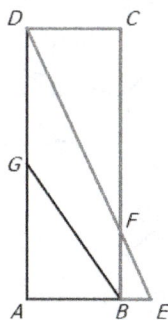

Because $CD//AB$, $\triangle CDF \sim \triangle BEF$. Thus, $CF/FB = CD/BE = 6/2 = 3$. So $CF = AD * (3/4) = 45/2$.

271

Therefore, the area of triangle CDF and ABG is equal to $135/2$ and 45. The area of $BFDG$ is equal to $180 - 135/2 - 45 = 135/2$.

The answer is (C).

Solution 2:

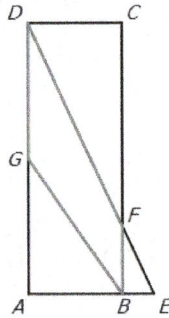

Because $CD//AB$, $\triangle CDF \sim \triangle BEF$. Thus, $CF/FB = CD/BE = 6/2 = 3$.

$FB = AD/4 = 15/2$.

The area of trapezoid $BFDG$ is equal to $(FB + DG) * (AB/2) = (15 + 15/2) * (6/2) = 135/2$.

The answer is (C).

Problem 187

A solid tetrahedron is sliced off a wooden unit cube by a plane passing through two nonadjacent vertices on one face and one vertex on the opposite face not adjacent to either of the first two vertices. The tetrahedron is discarded and the remaining portion of the cube is placed on a table with the cut surface face down. What is the height of this object? (2012 AMC10B)

(A) $\sqrt{3}/3$ (B) $2\sqrt{2}/3$ (C) 1   (D) $2\sqrt{3}/3$ (E) $\sqrt{2}$

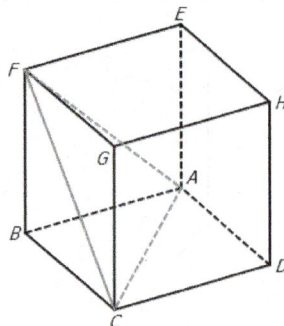

Tips: *1. Find the highest point to the base. 2. Calculate the altitude through volume method or cutting planes.*

Solution 1:

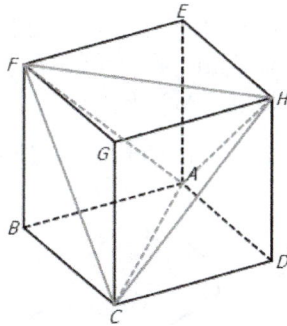

The height of the object is the distance between point H (diagram above) and triangle ACF. We can use volume method to find the distance between point H and plane ACF.

The volume of the tetrahedron sliced off ($ABCF$) is $1/6$. The tetrahedrons $ACDH$, $GCFH$, and $AFHE$ are congruent to the tetrahedron $ABCF$. Thus, the volume of tetrahedron $ACFH$ is $1 - (1/6) \times 4 = 1/3$.

Triangle ACF is an equilateral triangle with side-length of $\sqrt{2}$, so its area is equal to $\sqrt{3}/2$. Therefore, the distance between H and plane ACF is equal to $3V/S_{\triangle ACF} = 3 \times (1/3)/(\sqrt{3}/2) = 2\sqrt{3}/3$.

The answer is (D).

Solution 2:

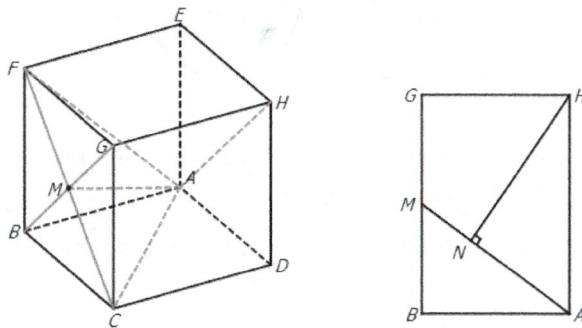

The height of the object is the distance between point H (diagram above) and triangle ACF.

Because $AB \perp FC$ and $BG \perp FC$, $FC \perp ABGH$. Let point N be on AM such that $HN \perp AM$. Together with $FC \perp HN$, we know that $HN \perp ACF$ and HN is the altitude line between point H and plane ACF. In the diagram, $BM = \sqrt{2}/2$, $MA = \sqrt{3/2}$, and $HA = \sqrt{2}$. Therefore, $HN = HA * AB/MA = \sqrt{2} * 1/\sqrt{3/2} = 2\sqrt{3}/3$.

The answer is (D).

273

Problem 188

Circles A, B, and C each have radius 1. Circles A and B share one point of tangency. Circle C has a point of tangency with the midpoint of AB. What is the area inside Circle C but outside circle A and circle B ? (2011 AMC10A)

(A) $3 - \pi/2$ (B) $\pi/3$ (C) 2 (D) $3\pi/4$ (E) $1 + \pi/2$

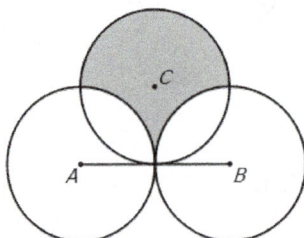

Tips: *Connect center of the three circles to the intersection points.*

Solution 1:

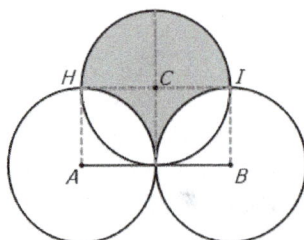

Connect AH, BI, and HI as shown in the diagram. The area of the semicircle above segment HI is equal to the area of the two quarter circles below segment HI in rectangle $ABIH$.

Therefore, the shaded area is equal to the area of rectangle $ABIH$, whose area is 2.

The answer is (C).

Solution 2:

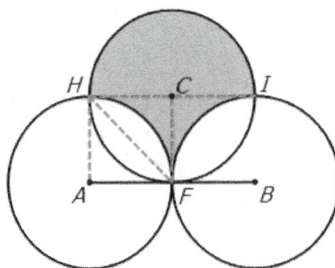

The shaded region is divided into three pieces as shown in the diagram. The top part is a semicircle and the two corner parts are below segment HI.

274

The area of the corner part is equal to the area of the unit square $AFCH$ subtracted by the area of a quarter circle. Thus, the area of corner part is equal to $1 - \pi/4$.

Therefore, the area of the shaded region is $\pi/2 + 2*(1 - \pi/4) = 2$.

The answer is (C).

Problem 189

In the given circle, the diameter EB is parallel to DC, and AB is parallel to ED. The $\angle AEB$ and $\angle ABE$ are in the ratio $4:5$. What is the degree measure of $\angle BCD$? (2011 AMC10B)

(A) 120   (B) 125   (C) 130   (D) 135   (E) 140

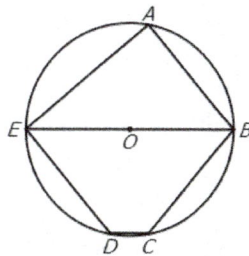

Tips: *Parallel lines leads to equal angles and equal arcs.*

Solution 1:

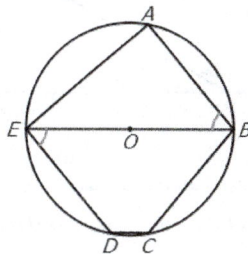

Because EB is the diameter of the circle, $\angle BAE = 90°$. Given $\angle AEB : \angle ABE = 4:5$, we find $\angle AEB = 40°$ and $\angle ABE = 50°$.

Because $AB//DE$ and $CD//BE$, $\angle BED = \angle ABE = \angle EBC = 50°$. Points B, C, D, and E are concyclic, so $\angle EDC = 180° - \angle EBC = 180° - 50° = 130°$.

The answer is (C).

Solution 2:

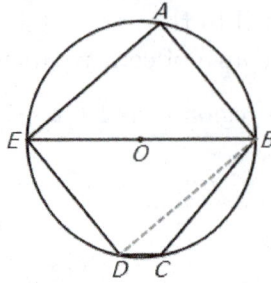

Connect BD. Because EB is the diameter of the circle, $\angle BAE = \angle BDE = 90°$. Given $\angle AEB : \angle ABE = 4 : 5$, we have $\angle AEB = 40°$ and $\angle ABE = 50°$. Because $AB//DE$ and $\angle BAE = \angle BDE = 90°$, $ABDE$ is a rectangle.

Because $CD//BE$, $\angle BDC = \angle DBE = \angle AEB = 40°$. Therefore, $\angle EDC = \angle EDB + \angle BDC = 90° + 40° = 130°$.

The answer is (C).

Problem 190

Rectangle $ABCD$ has $AB = 6$ and $BC = 3$. Point M is chosen on side AB so that $\angle AMD = \angle CMD$. What is the degree measure of $\angle AMD$? (2011 AMC10B)

(A) 15　　(B) 30　　(C) 45　　(D) 60　　(E) 75

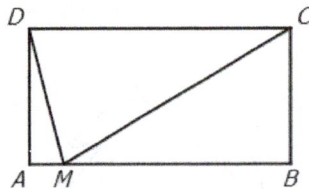

Tips: *Prove that triangle CMD is an isosceles triangle.*

Solution:

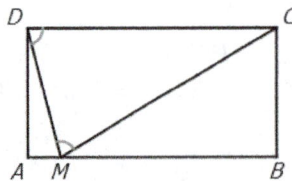

Because $AB//CD$, $\angle AMD = \angle CDM$. Given $\angle AMD = \angle CMD$, we know $\angle CDM = \angle CMD$. Thus, triangle CMD is an isosceles triangle, so $CM = CD = 6$.

276

In the right triangle CBM, $BC = CM/2$. Therefore, $\angle CMB = 30°$ and $\angle AMD = (180° - 30°)/2 = 75°$.

The answer is (E).

Problem 191

Rhombus $ABCD$ has side length 2 and $\angle B = 120°$. Region R consists of all points inside the rhombus that are closer to vertex B than any of the other three vertices. What is the area of R? (2011 AMC10B)

(A) $\sqrt{3}/3$ (B) $\sqrt{3}/2$ (C) $2\sqrt{3}/3$ (D) $1 + \sqrt{3}/3$ (E) 2

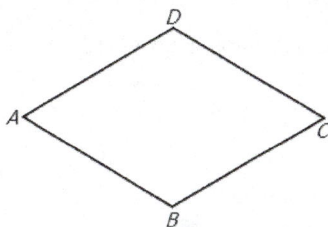

Tips: *To find the region of points that are closer to point B than to Point A, draw the perpendicular bisector of AB.*

Solution 1:

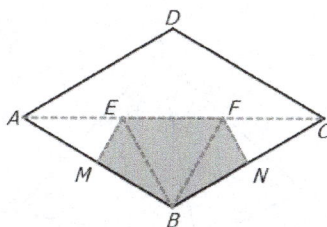

The targeted region is bounded by area between the perpendicular bisectors AB, BD, and BC as shown in the diagram. Any point in the shaded region is closer to vertex B than the other three points.

The area of the shaded region is equal to twice the area of equilateral triangle BEF, which has the area of $\sqrt{3}/3$. The total area is $2\sqrt{3}/3$.

The answer is (C).

Solution 2:

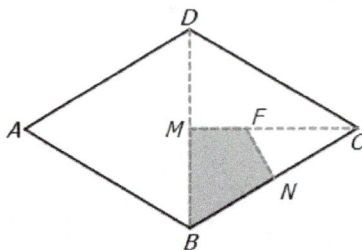

The geometric region is symmetric across the segment MB, so we need to find the area of the region in equilateral triangle BCD first. FN and MC are the perpendicular bisectors of BC and BD, respectively.

$BN = 1$ and $FN = \sqrt{3}/3$. The area of the shaded region is equal to the area of an equilateral triangle with a side length of $2\sqrt{3}/3$, which is $\sqrt{3}/3$.

The total area where the points are closer to vertex B than other three vertices is $2\sqrt{3}/3$.

The answer is (C).

Problem 192

A pyramid has a square base with sides of length 1 and has lateral faces that are equilateral triangles. A cube is placed within the pyramid so that one face is on the base of the pyramid and its opposite face has all its edges on the lateral faces of the pyramid. What is the volume of this cube? (2011 AMC10B)

(A) $5\sqrt{2} - 7$ (B) $7 - 4\sqrt{3}$ (C) $2\sqrt{2}/27$ (D) $\sqrt{2}/9$ (E) $\sqrt{3}/9$

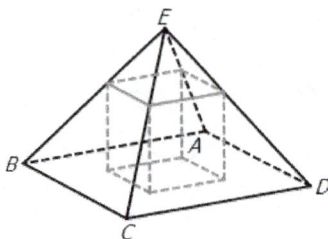

Tips: *Construct a cutting plane through the medians of the lateral faces to form a triangle with an inscribed square.*

Solution 1:

278

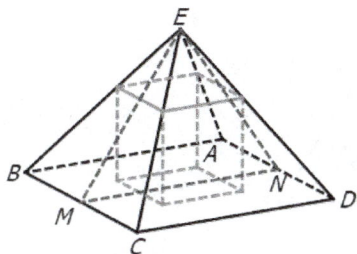

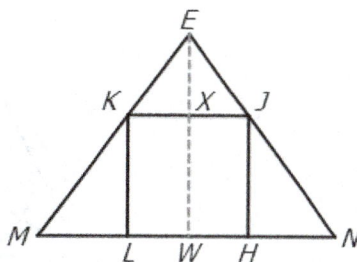

Let EM and EN be medians in the equilateral triangles EBC and EAD, respectively. We know that $EM = EN = \sqrt{3}/2$ and $MN = 1$. Cutting through EM and EN yields triangle EMN (right) with inscribed square $KJHL$. The altitude to MN is equal to $\sqrt{2}/2$.

Let $x = KJ$. The similar triangle pair EKJ and EMN yields $KJ/MN = EX/EW$. We get $x/1 = (\sqrt{2}/2 - x)/(\sqrt{2}/2)$. Solving the equation yields $x = \sqrt{2} - 1$. Therefore, the volume of the inscribed cube is equal to $(\sqrt{2} - 1)^3 = 5\sqrt{2} - 7$.

The answer is (A).

Solution 2:

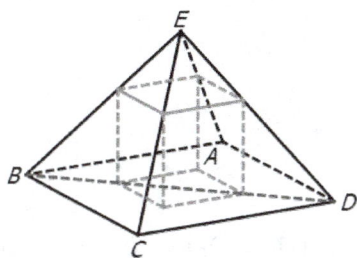

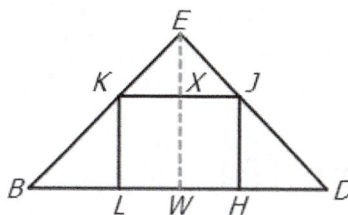

Cutting through points E, B and D yields triangle EBD with inscribed rectangle $KJHL$ and $KJ = \sqrt{2}KL$. We also know that $EB = 1$, $BD = \sqrt{2}$, and $EW = \sqrt{2}/2$.

Let $x = KL$. Because $\Delta EBD \sim \Delta EKJ$, $KJ/BD = EX/EW$ and $\sqrt{2}x/\sqrt{2} = (\sqrt{2}/2 - x)/(\sqrt{2}/2)$. Solving the equation gives us $x = \sqrt{2} - 1$, so the volume of the inscribed cube is equal to $(\sqrt{2} - 1)^3 = 5\sqrt{2} - 7$.

The answer is (A).

Problem 193

Let T_1 be a triangle with sides 2011, 2012, and 2013. For $n \geq 1$, if $T_n = \Delta ABC$ and D, E, and F are the points of tangency of the incircle of ΔABC to the sides AB, BC and AC, respectively, then T_{n+1} is a triangle with side lengths AD, BE, and CF, if it exists. What is the perimeter of the last triangle in the sequence T_n ? (2011 AMC10B)

(A) 1509/8   (B) 1509/32   (C) 1509/64   (D) 1509/128   (E) 1509/256

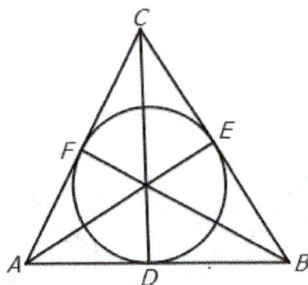

Tips: *Find a formula for the 3 edge lengths and apply the Triangle Inequality Theorem.*

Solution:

Let $b = 2012$. In T_1, the edge-lengths of AB, BC and AC are $b-1$, b and $b+1$, respectively. By applying the inscribed circle rules, $AD = b/2$, $BE = b/2 - 1$ and $CF = b/2 + 1$. Reordering the lengths, we have $b/2 - 1$, $b/2$ and $b/2 + 1$.

Similarly, the nth triangle has edge lengths of $b/2^{(n-1)} - 1$, $b/2^{(n-1)}$ and $b/2^{(n-1)} + 1$. The requirement that the three sides form a triangle gives us the inequality $(b/2^{(n-1)} - 1 + b/2^{(n-1)}) > b/2^{(n-1)} + 1$ and $b = 2012$. Solving the inequality gives $2^{(n-1)} < 2012/2$, so $n - 1 \le 9$. The last triangle has the perimeter of $3b/2^9 = 1509/128$.

The answer is (D).

Problem 194

Equiangular hexagon $ABCDEF$ has side lengths $AB = CD = EF = 1$ and $BC = DE = FA = r$. The area of $\triangle ACE$ is 70% of the area of the hexagon. What is the sum of all possible values of r? (2010 AMC10A)

(A) $4\sqrt{3}/3$ (B) $10/3$ (C) 4 (D) $17/4$ (E) 6

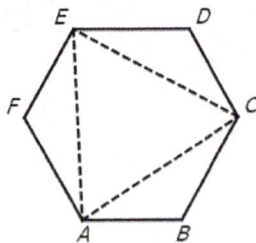

Tips: *Use the law of sines and the law of cosines to calculate the area of each triangle.*

Solution 1:

In an equiangular hexagon, each angle is equal to $120°$. The three triangle ABC, CDE and EFA are congruent, thus $AC = CE = EA$.

Because the area of $\triangle ACE$ is 70% of the area of the hexagon, the area of triangle ABC is 10% of the area of the hexagon. The area of triangle ABC is equal to $(1/2)*r*\sin 120° = (\sqrt{3}/4)*r$. By the law of cosines, $AC^2 = 1 + r^2 - 2*1*r*\cos 120° = r^2 + r + 1$.

The area of equilateral triangle ACE is equal to $(\sqrt{3}/4) * AC^2 = (\sqrt{3}/4) * (r^2 + r + 1)$. The area ratio between triangle ACE and ABC is 7, thus $7r = r^2 + r + 1$ and $r^2 - 6r + 1 = 0$. Using **Vieta's Theorem**, we know the sum of the possible values of r is 6.

The answer is (E).

Solution 2:

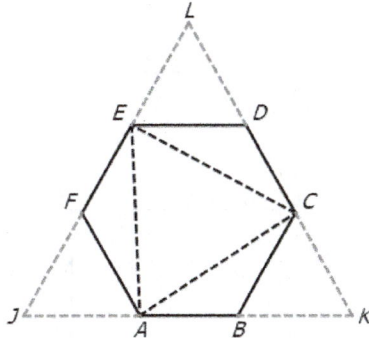

Extend AB, CD and EF to intersect at point K, L and J as shown in the diagram. The resulting $\triangle BCK$, $\triangle DEL$ and $\triangle AFJ$ are congruent equilateral triangle and $\triangle JKL$ is equilateral triangle with side-length of $1 + 2r$.

Because the area of triangle ACE is 70% of the area of the hexagon and triangle ABC, CDE and EFA are congruent, the area of triangle ABC is 10% of the area of the hexagon. The area of $\triangle BCK$ is 10% of the area of the hexagon and the area of $\triangle JKL$ is equal to $(1 + 30\%r)$ of the area of the hexagon.

Because $\triangle BCK \sim \triangle JKL$, the area ratio of the two triangle is equal to $(r/(1 + 2r))^2 = (10\%r)/(1 + 30\%r)$. Simplifying the equation gives $r^2 - 6r + 1 = 0$. Thus, the sum of the possible values of r is 6.

The answer is (E).

Problem 195

A square of side length 1 and a circle of radius $\sqrt{3}/3$ share the same center. What is the area inside the circle, but outside the square? (2010 AMC10B)

(A) $\pi/3 - 1$ (B) $2\pi/9 - \sqrt{3}/3$ (C) $\pi/18$ (D) $1/4$ (E) $2\pi/9$

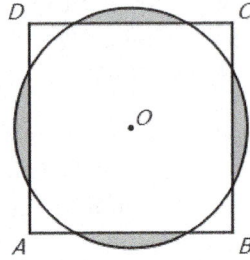

Tips: *Connect center of the circle to the intersecting points with square to form a right triangle.*

Solution:

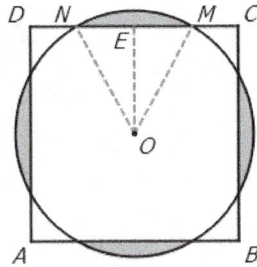

Let points M and N be the intersecting points and connect OM, ON and median OE. Because $OE = 1/2$, $OM = ON = \sqrt{3}/3$.

In right triangle OEM, by **Pythagorean Theorem**, $MN = 2EM = 2\sqrt{(\sqrt{3}/3)^2 - (1/2)^2} = \sqrt{3}/3$. Thus, $\triangle OMN$ is an equilateral triangle. The area of $\triangle OMN$ is equal to $(\sqrt{3}/4) * (\sqrt{3}/3)^2 = \sqrt{3}/12$ and the area of $60°$ pie is equal to $(1/6\pi) * (\sqrt{3}/3)^2 = \pi/18$.

Therefore, the area of the shaded region is equal to $4(\pi/18 - \sqrt{3}/12) = 2\pi/9 - \sqrt{3}/3$.

The answer is (B).

Problem 196

A circle with center O has area 156π. Triangle ABC is equilateral, BC is a chord on the circle, $OA = 4\sqrt{3}$, and point O is outside $\triangle ABC$. What is the side length of $\triangle ABC$? (2010 AMC10B)

(A) $2\sqrt{3}$ (B) 6 (C) $4\sqrt{3}$ (D) 12 (E) 18

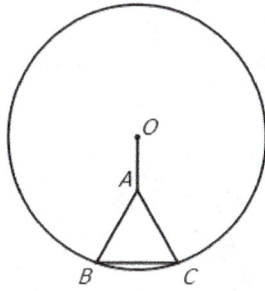

Tips: *Connect OB and OC to form a triangle and draw the altitude of the triangle.*

Solution:

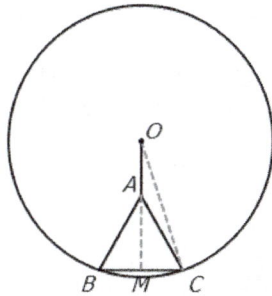

Let point M be the midpoint of segment BC. Connect OC and draw the altitude AM of the equilateral $\triangle ABC$. The radius of the circle O is equal to $\sqrt{156}$.

Let $BC = 2x$. Thus, $OM = OA + AM = 4\sqrt{3} + \sqrt{3}x$. In right triangle OMC, by the **Pythagorean Theorem**, $x^2 + (4\sqrt{3} + \sqrt{3}x)^2 = 156$. Solving the equation yields $x = 3$, so the side length of $\triangle ABC$ is 6.

The answer is (B).

Problem 197

Regular octagon $ABCDEFGH$ has area n. Let m be the area of quadrilateral $ACEG$. What is m/n ? (2020 AMC12A)

(A) $\sqrt{2}/4$ (B) $\sqrt{2}/2$ (C) $3/4$ (D) $3\sqrt{2}/5$ (E) $2\sqrt{2}/3$

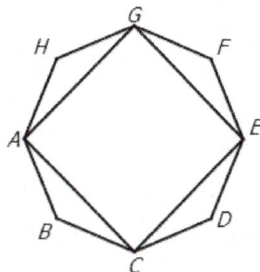

Tips: *Connect the center to the sides of octagon and the quadrilateral to form triangles.*

Solution 1:

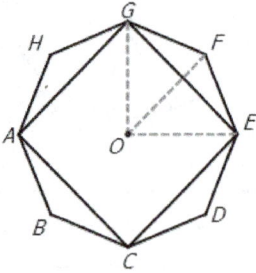

Let the radius of the circumcircle be r. Thus, $OG = OE = OF = r$. The area of quadrilateral $ACEG$ (square) is equal to $2r^2$.

For a regular octagon, $\angle GOF = 45°$. The area of triangle GOF is $(1/2) * r^2 \sin 45° = r^2\sqrt{2}/4$ and the area of $ABCDEFGH$ is $r^2\sqrt{2}/4 * 8 = 2\sqrt{2}r^2$.

Therefore, $m/n = (2r^2)/(2\sqrt{2}r^2) = \sqrt{2}/2$.

The answer is (B).

Solution 2:

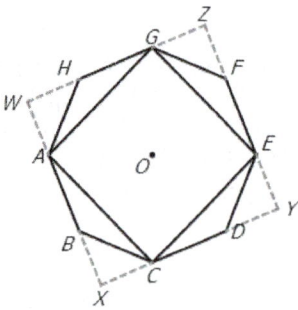

Extend segment AB, CD, FE, and GH to intersect at point X, Y, Z and W. Quadrilateral $XYZW$ is a square and $\triangle XBC$, $\triangle YDE$, $\triangle ZFG$, and $\triangle WHA$ are congruent isosceles right triangles.

Let $x = GZ$. Thus, $EF = \sqrt{2}x$, $ZE = (1+\sqrt{2})x$, $ZY = (2+\sqrt{2})x$, and $GE^2 = (4+2\sqrt{2})x^2$, which is the area of quadrilateral $ACEG$. The area of the octagon $ABCDEFGH$ is equal to $(2+\sqrt{2})^2x^2 - 2x^2 = (4+4\sqrt{2})x^2$. Thus, $m/n = ((4+2\sqrt{2})x^2)/((4+4\sqrt{2})x^2) = \sqrt{2}/2$.

The answer is (B).

Problem 198

Suppose that $\triangle ABC$ is an equilateral triangle of side length s, with the property that there is a unique point P inside the triangle such that $AP = 1$, $BP = \sqrt{3}$, and $CP = 2$. What is s? (2020 AMC12A)

284

(A) $1+\sqrt{2}$ (B) $\sqrt{7}$ (C) $8/3$ (D) $\sqrt{5+\sqrt{5}}$ (E) $2\sqrt{2}$

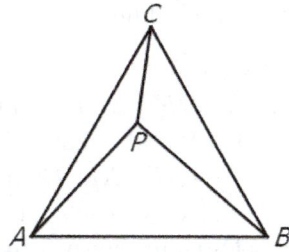

> **Tips:** *1. Given that the lengths form a Pythagorean triplet, you want to put them into a single triangle. 2. Rotate one of the segments connected to P by 60 degrees to create another equilateral triangle.*

Solution 1:

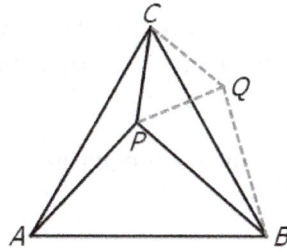

Rotate segment PC around vertex C by $60°$ to QC as shown in the diagram. The resulting triangle PQC is an equilateral triangle. In the $\triangle PQC$, $PQ = CQ = PC = 1$ and $\angle PCQ = 60°$.

Because $\angle ACB = \angle PCQ = 60°$, $\angle QCB = \angle PCA$. Thus, $\triangle BCQ \cong \triangle ACP$ and $QB = PA = \sqrt{3}$. In the triangle PQB, $PB^2 = PQ^2 + QB^2$. By the converse of the **Pythagorean Theorem**, $\angle PQB = 90°$, so $\angle BQC = 90° + 60° = 150°$.

By the law of cosines, $BC^2 = 1 + 3 - 2\sqrt{3} * \cos 150° = 7$. Therefore, $s = BC = \sqrt{7}$.

The answer is (B).

Solution 2:

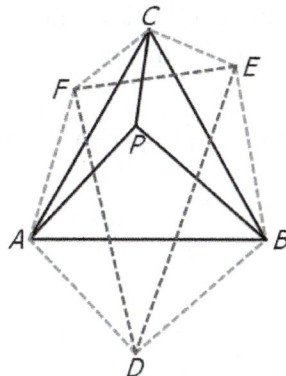

285

Reflect point P over segments AB, BC and AC to point D, E and F as shown in the diagram and connect CE, EF and FD. With the reflection, we know $\triangle ABP \cong \triangle ABD$, $\triangle BCP \cong \triangle BCE$ and $\triangle ACP \cong \triangle ACF$. The area of $ADBECF$ is twice the area of triangle ABC.

It is straightforward to find that $\triangle ADF$, $\triangle BDE$ and $\triangle CEF$ are isosceles triangles and $\angle DAF = \angle DBE = \angle ECF = 120°$. We also have $AD = AF = \sqrt{3}$, $BD = BE = 2$ and $CE = CF = 1$. Thus, the area of $\triangle ADF$, $\triangle BDE$ and $\triangle CEF$ are $3\sqrt{3}/4$, $\sqrt{3}/4$ and $\sqrt{3}$, respectively. Applying the law of cosines in each triangle gives $DF = 3$, $DE = 2\sqrt{3}$, and $FE = \sqrt{3}$. By the converse of the **Pythagorean Theorem**, $\angle DFE = 90°$. Thus, the area of triangle DEF is equal to $3\sqrt{3}/2$.

Therefore, the area of $ADBECF$ is $7\sqrt{3}/2$ and the area of triangle ABC is $7\sqrt{3}/4$. Based on the area formula for an equilateral triangle, the side-length of an equilateral triangle with an area of $7\sqrt{3}/4$ is $\sqrt{7}$.

The answer is (B).

Solution 3:

We can use analytic geometry to solve this problem. Assume $s = 2a$ and let the coordinates of vertices A, B, and C be $(-a, 0)$, $(a, 0)$, and $(0, \sqrt{3}a)$, respectively. Let the coordinate of point P be (x, y).

The distance formula gives us the following equations:

$(x + a)^2 + y^2 = 3$ (1)

$(x - a)^2 + y^2 = 4$ (2)

$x^2 + (y - \sqrt{3}a)^2 = 1$ (3)

(1) - (2) gives $4ax = -1$ and $x = -1/(4a)$. Substituting x into (1) and simplifying the equation gives $y = \sqrt{7/2 - a^2 - 1/(16a^2)}$. Substituting both x and y into (3) and solving the equation gives $a = \sqrt{7}/2$. Therefore, $s = 2a = \sqrt{7}$.

The answer is (B).

Problem 199

As shown in the figure below, six semicircles lie in the interior of a regular hexagon with side length 2 so that the diameters of the semicircles coincide with the sides of the hexagon. What is the area of the shaded region — inside the hexagon but outside all of the semicircles? (2020 AMC12B)

(A) $6\sqrt{3} - 3\pi$ (B) $9\sqrt{3}/2 - 2\pi$ (C) $3\sqrt{3}/2 - \pi/3$ (D) $3\sqrt{3} - \pi$ (E) $9\sqrt{3}/2 - \pi$

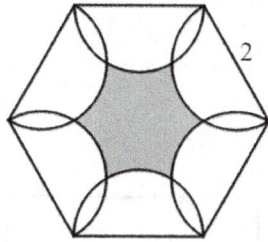

Tips: *Connect the centers to the points of intersection.*

Solution:

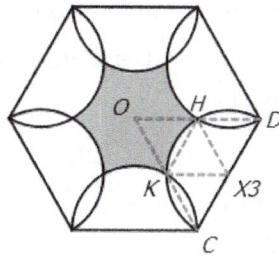

Connect the center O and two adjacent vertices C and D to form an equilateral triangle OCD (isosceles triangle OCD with $\angle ODC = 60°$). Connect the intersection point H and center of the semicircle X_3 to form an smaller equilateral triangle DHX_3. The total shaded area in the triangle OCD is equal to the area difference of 12 small equilateral triangles and one full circle.

We know $OH = OK = CX_3 = 1$, so the total area of the small equilateral triangles is $12 * \sqrt{3}/4 = 3\sqrt{3}$. Therefore, the shaded area is $3\sqrt{3} - \pi$.

The answer is (D).

Problem 200

Rectangle ABCD is shown below with $CD = 10$ and $AD = 8$. If E is the midpoint of BC , G is the midpoint of AB, F is a point on CD, and the area of EFG is 18, find the length of DF.

(A) 4   (B) 5   (C) 6   (D) 7   (E) 8

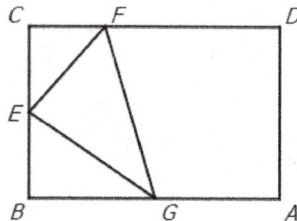

Solution:

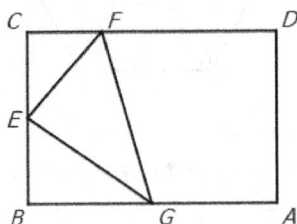

We know that the total area is 80, so the area excluding the triangle is $80 - 18 = 62$.

Let $FD = x$. $FC = 10 - x$, $CE = BE = 4$, and $BG = AG = 5$.

$S_{\triangle BEG} + S_{\triangle CEF} + S_{ADFG} = 62$. (1)

$S_{\triangle BEG} = (1/2) * 5 * 4 = 10$.

$S_{\triangle CEF} = (1/2) * (10 - x) * 4 = 2(10 - x)$.

$S_{ADFG} = (1/2) * (5 + x) * 8 = 4(x + 5)$

Substituting into equation (1) gives:

$10 + 2(10 - x) + 4(x + 5) = 62$

Solving the equation gives us $x = 6 = DF$.

The answer is (C).

Problem 201

Circles ω and γ, both centered at O, have radii 20 and 17, respectively. Equilateral triangle ABC, whose interior lies in the interior of ω but in the exterior of γ, has vertex A on ω, and the line containing side BC is tangent to γ. Segments AO and BC intersect at P, and $BP/CP = 3$. Then AB can be written in the form $m/\sqrt{n} - p/\sqrt{q}$ for positive integers m, n, p, q with $\gcd(m,n) = \gcd(p,q) = 1$. What is $m + n + p + q$? (2019 AMC12A)

(A) 42   (B) 86   (C) 92   (D) 114   (E) 130

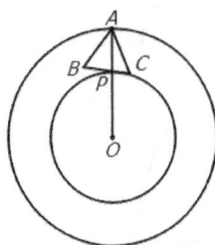

Tips: *1. Connect the center of the circle to the tangent points and construct the altitude of $\triangle ABC$.*

Solution:

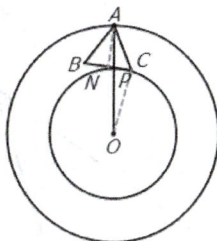

Connect O to the tangent point P and let AN be the median of triangle ABC. Thus $ON = 17$.

Let AB be $4a$. Therefore, $AN = 2\sqrt{3}a$ and $NP = (3/4 - 1/2) * (4a) = a$. Because $\triangle ANP \sim \triangle OCP$ and $OA = 20$, $AP = OA*AN/(AN+ON) = 20*2\sqrt{3}a/(17+2\sqrt{3}a) = 40\sqrt{3}a/(17+2\sqrt{3}a)$.

In the right triangle ANP, by the **Pythagorean Theorem**, $AP^2 = AN^2 + NP^2$, so $40\sqrt{3}a/(17 + 2\sqrt{3}a) = \sqrt{13}a$. Solving the equation yields $4a = 80/\sqrt{13} - 34/\sqrt{3}$. Therefore, $m + n + p + q = 130$.

The answer is (E).

Problem 202

In $\triangle ABC$ with integer side lengths,

$$\cos A = 11/16, \cos B = 7/8, \text{ and } \cos C = -1/4$$

What is the least possible perimeter for $\triangle ABC$? (2019 AMC12A)

(A) 9   (B) 12   (C) 23   (D) 27   (E) 44

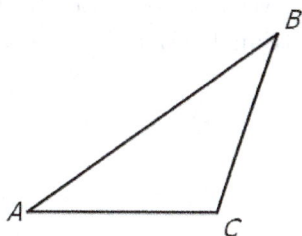

Tips: *Use law of sines and law of cosines to calculate the ratio of edge-lengths.*

Solution 1:

Because $\cos A = 11/16$, $\cos B = 7/8$, and $\cos C = -1/4$, $\sin A = 3\sqrt{15}/16$, we have $\sin B = \sqrt{15}/8$, and $\sin C = \sqrt{15}/4$. By the law of sines, $a : b : c = \sin A : \sin B : \sin C = 3 : 2 : 4$. Therefore, the least possible perimeter for triangle ABC with integer lengths is 9.

Answer is (A).

Solution 2:

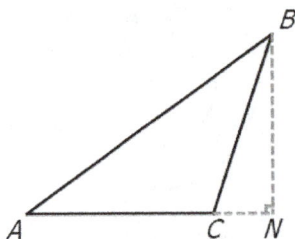

Construct altitude BN to side AC as shown in the diagram. Let $BC = 4x$. Because $\cos C = -1/4$, $CN = x$ and $BN = \sqrt{15}x$. $\sin A = \sqrt{1 - (\cos A)^2} = 3\sqrt{15}/16$.

$c = AB = BN/\sin A = (16/3) * x$. $b = AC = AN - CN = AB \cos A - x = (11/3) * x - x = (8/3) * x$.

Therefore, $a : b : c = 4x : (16/3) * x : (8/3) * x = 3 : 4 : 2$. The least possible perimeter for triangle ABC is 9.

The answer is (A).

Solution 3:

By the law of cosines, we have

$a^2 + b^2 - c^2 = 2ab \cos C = -ab/2$ (1)

$b^2 + c^2 - a^2 = 2bc \cos A = 11bc/8$ (2)

$c^2 + a^2 - b^2 = 2ac \cos B = 7ac/4$ (3)

(1) + (2) gives $2b = 11/8c - a/2$

(1) + (3) gives $2a = 7/4c - b/2$

Solving the equation set above gives $a = (3/4) * c$ and $b = c/2$. Thus, $a : b : c = 3 : 4 : 2$ and the least possible perimeter for triangle ABC is 9.

The answer is (A).

Problem 203

In triangle ABC, $AB = 10$, $AC = 17$, and $BC = 21$. What is the area of the circumscribed circle. (Mathcounts 2013, National Sprint)

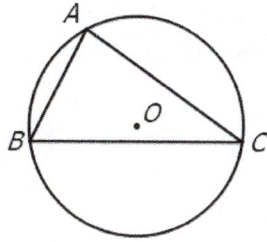

Tips: *1. The goal is to find the radius/diameter of the circumscribed circle. 2. Use area formula involving radius or law of sines to calculate the radius.*

Solution 1:

The first step is to use **Heron's Formula** to calculate the area of $\triangle ABC$.

$s = (AB + BC + CA)/2 = (10 + 17 + 21)/2 = 24$. **Heron's formula** yields the area $A = \sqrt{s(s-a)(s-b)(s-c)} = 84$.

The area of $\triangle ABC$ can be expressed as $A = AB * BC * CA/(4r) = (10 * 17 * 21)/(4r)$. Thus, $r = 85/8$ and the area of $\triangle ABC$ is $7225\pi/64$.

Solution 2:

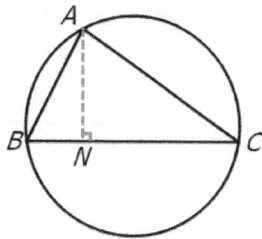

Instead of using Heron's Formula to calculate the area of $\triangle ABC$, we want to find the altitude on base BC. N is the point on BC such that $AN \perp BC$. Letting $BN = x$, $CN = y$ and $AN = h$, we have $x + y = BC = 21$.

Applying the **Pythagorean Theorem** in $\triangle ABN$ and $\triangle ANC$ yields $AC^2 - AB^2 = (y - x)(y + x) = 17^2 - 10^2$. Solving the equation set yields $y = 15$ and $x = 6$. Therefore, $h = 8$ and the area of $\triangle ABC$ is equal to 84.

The area of $\triangle ABC$ can be expressed as $A = AB * BC * CA/(4 * r) = (10 * 17 * 21)/(4r)$. Thus, $r = 85/8$ and the area of $\triangle ABC$ is $7225\pi/64$.

Solution 3:

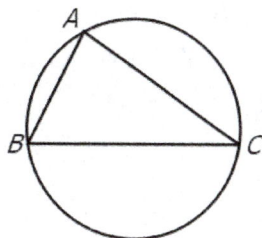

In this solution, we use the law of sines to calculate the radius of the circumscribed circle. Knowing the edge-length, the law of cosines yields $\cos B = (AB^2 + BC^2 - AC^2)/(2AB*BC) = 3/5$. Thus, $\sin B = 4/5$.

The law of sines yields $r = AC/2 * \sin B = 85/8$. Therefore, the area of $\triangle ABC$ is $7225\pi/64$.

Problem 204

Trapezoid $KLMN$ has sides $KL = 80$ units, $LM = 60$ units, $MN = 22$ units, and $KN = 65$ units, with KL parallel to MN. A semicircle with center A on KL is drawn tangent to both sides KN and ML. What is the length of segment KA? (Mathcounts 2013, National Sprint)

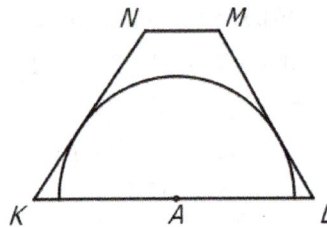

Tips: *Semicircle being tangent to both sides indicates that A is on the angular bisector of the extended triangle.*

Solution 1:

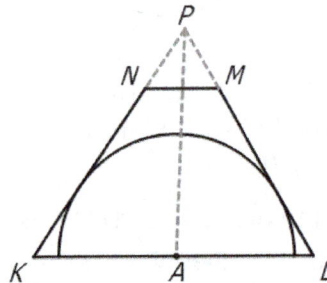

Extend KN and LM to meet at point P and connect PA. Because the semicircle is tangent to both KN and ML, PA is the angular bisector of $\angle KPL$.

The **Angular Bisector Theorem** yields $KA/AL = PK/PL$. Because $MN//KL$, $PK/PL = NK/ML$. Therefore, $KA/AL = NK/ML = 65/60$.

Given that $KL = KA + AL = 80$, applying the **Angular Bisector Theorem** gives us

$KA = 80 \times 65/(65 + 60) = 208/5$.

Solution 2:

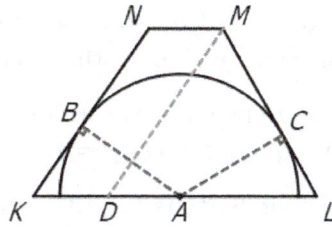

Let point D be on KL such that $MD//NK$. Connect A to the tangent points B and C.

Because $MN//KL$ and $MD//NK$, $MD = KN = 65$ and $\angle MDL = \angle K$. In the $\triangle MDL$, the law of sines yields $\sin \angle MDL / \sin L = ML/MD = 60/65$. In the $Rt\triangle ABK$ and $Rt\triangle ACL$, $AB = AC = KA \sin K = AL \sin L$. Thus, $KA/AL = \sin L / \sin K = \sin \angle MDL / \sin L = 60/65$.

Given $KA + AL = 80$, $KA = 80 \times 65/(65 + 60) = 208/5$.

Problem 205

A circle is tangent to the positive x-axis at $x = 3$. It passes through the distinct points $(6, 6)$ and (p, p). What is the value of p? Express your answer as a common fraction. (Mathcounts 2020 State Sprint)

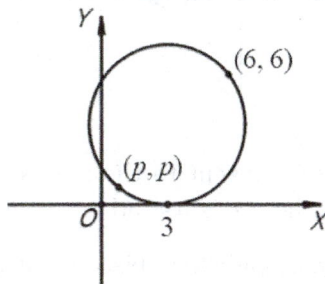

Tips: *1. Establish the equation of a circle centered at (3,r) with a radius of r. 2. Substitute the other two points on the circle to calculate r and p.*

Solution 1:

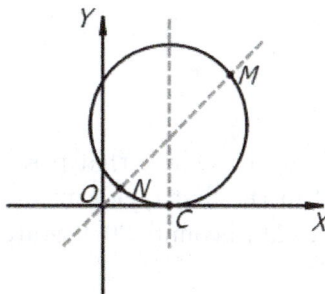

293

The circle is tangent to the x-axis at $C(3,0)$, so the center of the circle can be represented by $(3,r)$, where r is its radius. The general form of the circle is $(x-3)^2 + (y-r)^2 = r^2$.

Because point $M(6,6)$ is on the circle, substituting its coordinate into the circle yields $(6-3)^2 + (6-r)^2 = r^2$. Solving the equation gives $r = 15/4$.

The point $N(p,p)$ is also on the circle, thus $(p-3)^2 + (p-r)^2 = r^2$. Solving the equation gives $p = 3/4$ or $p = 6$ (point M). The answer is $p = 3/4$.

Solution 2:

Similar to solution 1, the circle is tangent to the x-axis at $C(3,0)$, so the center of the circle can be represented by $P(3,r)$, where r is its radius.

Points M and N are on the circle, so $PM = PN = r$. The distance formula for $PM = r$ yields $\sqrt{(6-3)^2 + (6-r)^2} = r$. Solving the equation yields $r = 15/4$. Similarly, the distance for $PN = r$ yields $\sqrt{(p-3)^2 + (p-r)^2} = r$. Solving the equation yields $p = 3/4$ or $p = 6$ (point M). The answer is $p = 3/4$.

Solution 3:

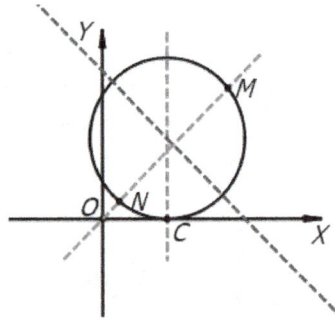

Similar to solution 1, the circle is tangent to the x-axis at $C(3,0)$, so the center of the circle can be represented by $P(3,r)$, where r is its radius.

The center P also lies on the perpendicular bisector of MN. The midpoint of MN is $(3 + p/2, 3 + p/2)$ and the slope of the perpendicular bisector is -1. The the equation of the perpendicular bisector is $y = -x + 6 + p$. When $x = 3$, $y = p + 3 = r$.

Similarly, the center P is also on the perpendicular bisector of MC. The midpoint of MC is $(9/2, 3)$ and the slope of MC is 2. Thus, the equation of the perpendicular bisector is $y = -(1/2) * (x - 9/2) + 3$. When $x = 3$, $y = 15/4 = r$.

Therefore, $p = r - 3 = 15/4 - 3 = 3/4$.

Problem 206

What is the greatest possible radius of a circle that passes through the points $(1,2)$ and $(4,5)$ and whose interior is contained in the first quadrant of the coordinate plane? Express your answer in simplest radical form. (Mathcounts 2019 State Sprint)

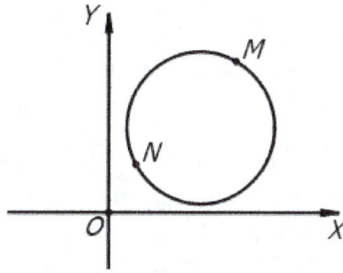

Tips: *1. The center of the circle is on the perpendicular bisector of the two known points. 2. The circle must be in the first quadrant so the circle is tangent to either the x or y-axis.*

Solution 1:

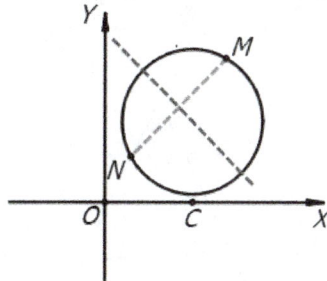

Because the circle passes through point M and N, the center of the circle is on the perpendicular bisector of MN. The midpoint of MN is $(5/2, 7/2)$ and the slope of MN is 1. Thus, the equation of the perpendicular bisector is $y - 7/2 = -(x - 5/2)$. Simplifying the equation gives us $y = -x + 6$.

Let the x-coordinate of the center be a. Then, $y = 6 - a$. The coordinates are $(a, 6 - a)$. The greatest radius occurs when the circle is tangent to one axis and not crossing the other axis. Two cases:

(1) The circle is tangent to the x-axis. Then, $r = 6 - a$ and $r \leq a$. Solving the inequality gives $r \leq 3$.

(2) The circle is tangent to the y-axis. Then, $r = a$ and $r \leq 6 - a$. Solving the inequality gives $r \leq 3$.

Therefore, the greatest radius of the circle is 3.

Solution 2:

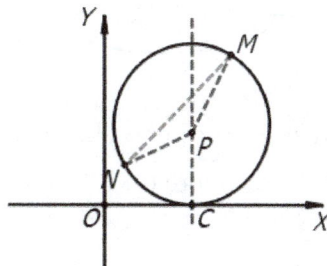

The circle with greatest radius is when the circle is tangent to one (or both) of the axis.

Case (1): The circle is tangent to the x-axis. Assuming the radius of the circle be r, the center of the circle can then be expressed as (a, r), where $r \leq a$.

Because point M and N are on the circle, the distance formula gives $(a-1)^2 + (r-2)^2 = (a-4)^2 + (r-5)^2$. Simplifying the equation gives $a + r = 6$. Combining with $r \leq a$ gives $r \leq 3$. When $r = 3$, $a = 3$. The circle is also tangent to the y-axis, so there is no need to discuss other cases. The greatest radius of the circle is 3.

Problem 207

A cone with base radius 12 cm is sliced parallel to its base, as shown, to remove a smaller cone of height 15 cm. If the height of the smaller cone is three-fourths that of the original cone, what is the volume of the remaining frustum? Express your answer in terms of π. (Mathcounts 2019 State Sprint)

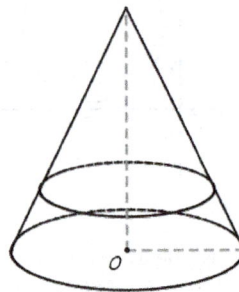

> **Tips:** *Use the properties of similar triangles to calculate the radius of the top face in the frustum.*

Solution 1:

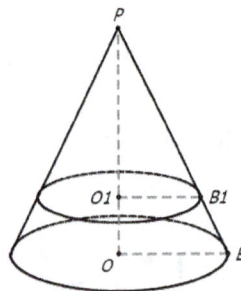

Because PO_1 is 3/4 of PO and $PO1 = 15$ cm, $PO = 20$ cm and $OO_1 = 5$ cm.

We also have $\triangle PO_1 B_1 \sim \triangle POB$, therefore $O1B1/OB = PO1/PO = 3/4OB = 3/4 \times 12 = 9$ cm.

The volume of the original big cone is $\pi h r^2/3 = 20 \times 12^2/3 \times \pi = 960\pi$. The volume of the small cone is $\pi h_1 r_1^2/3 = 15 \times 9^2/3 \times \pi = 405\pi$. Therefore, the volume of the frustum is 555π.

Solution 2:

Similar to solution #1, $PO_1 = 15$ cm, $PO = 20$ cm, and $OO_1 = 5$ cm. We also have $\Delta PO_1B_1 \sim \Delta POB$, so $O_1B_1/OB = PO_1/PO = 3/4 \times OB = 3/4 \times 12 = 9$ cm.

Applying the **Volume Formula** for a frustum yields $V = \pi h(a^2 + ab + b^2)/3 = \pi(144 + 108 + 81) \times 5/3 = 555\pi$.

Problem 208

The circles given by the equations $x^2 + y^2 = 169$ and $x^2 + (y - 14)^2 = 225$ have a common chord. How many units long is that chord? (Mathcounts 2019 State Sprint)

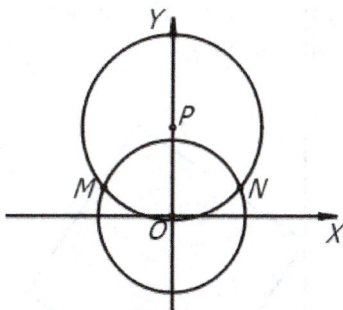

Tips: 1. Construct a triangle consisting of the centers of the two circles and the intersection point and calculate the altitude. 2. Solving the equation set to obtain the coordinates information of the two intersecting points.

Solution 1:

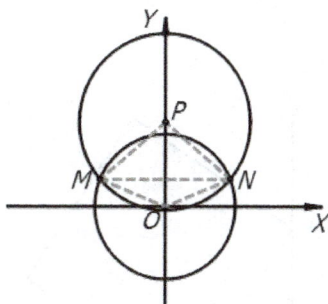

Points M and N are the intersecting points. Connect OM, ON, PM and PN. We have $PM = 15$, $OP = 14$, and $OM = 13$. The common cord MN is twice the length of the altitude on OP in triangle OMP.

Knowing the three edge-lengths, there are multiple ways to calculate its altitude. Here we use **Heron's Formula.** $s = (13 + 14 + 15)/2 = 21$, so the area $A = \sqrt{21 \times 6 \times 7 \times 8} = 84$.

297

Therefore, the length of the common chord is $2A/OP = 2 \times 84/14 = 12$.

Solution 2:

Solving the equation set to obtain the coordinates of the intersecting points:

$x^2 + y^2 = 169$ (1)

$x^2 + (y - 14)^2 = 225$ (2)

Equation (1) - (2) yields $28y = 169 - 225 + 196$. Thus, $y = 5$. Substituting it into equation (1) yields $x = \pm 12$. Therefore, the length of the common chord is 24.

Problem 209

Chloe finds a coin, as shown, wedged tightly into the corner of her drawer behind a rectangular box with a 2.5-inch edge. One corner of the box is 1.5 inches from the corner of the drawer and the other corner of the box is 2 inches from the corner of the drawer. What is the diameter of the coin? (Mathcounts 2017 State Sprint)

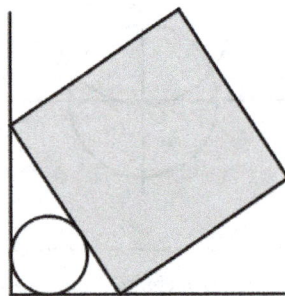

Tips: *1. Calculate the radius using the area of the triangle or the special formula for right triangles.*

Solution 1:

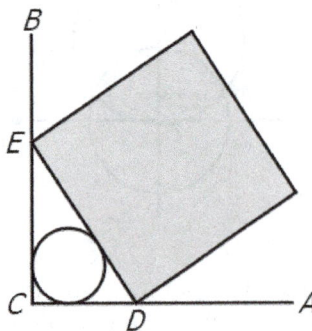

In the right triangle CDE, $CD = 1.5$ inches, $CE = 2$ inches, and $DE = 2.5$ inches. The diameter of the inscribed circle in a right triangle is equal to $CD + CE - DE = 1$ inch.

Solution 2:

In the right triangle CDE, $CD = 1.5$ inches, $CE = 2$ inches, and $DE = 2.5$ inches.

For any triangle, its area $A = rs$, where r is the radius of the inscribed circle and s is half of its perimeter.

For the right triangle CDE, $s = (1.5 + 2.5 + 2)/2 = 3$ and $A = 1.5 \times 2/2 = 1.5$. Therefore, $r = 1/2$ and the diameter is $2r = 1$ inch.

Problem 210

What is the radius of the inscribed circle of a triangle with side lengths 9, 13 and 14? Express your answer in simplest radical form. (Mathcounts 2017 State Sprint)

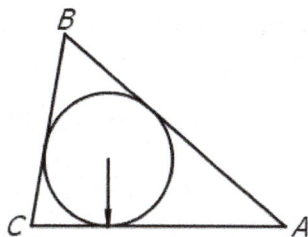

Tips: *Calculate the area of the triangle and then obtain the radius of the inscribed circle.*

Solution 1:

In $\triangle ABC$, let $AC = 14$, $AB = 13$ and $BC = 9$.

The semi-perimeter is $s = (14 + 13 + 9)/2 = 18$. Applying the **Heron's Formula** yields $A = \sqrt{18 \times 4 \times 5 \times 9} = 18\sqrt{10}$.

Therefore, $r = A/s = \sqrt{10}$.

Solution 2:

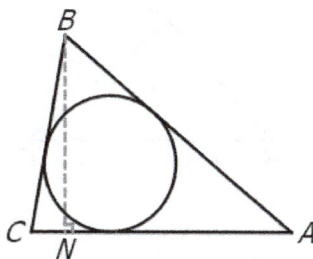

In $\triangle ABC$, let $AC = 14$, $AB = 13$, and $BC = 9$. The half perimeter is $s = (14+13+9)/2 = 18$. Find the altitude to calculate the area of the triangle.

Applying the **Pythagorean Theorem**, $AB^2 - BC^2 = (AN - CN) * AC$. Thus, $AN - CN = 44/7$. Combined with the equation $AN + CN = 14$, we have $AN = 71/7$. Then, $BN = 18\sqrt{10}/7$.

The area of the triangle ABC is equal to $BN * AC/2 = 18\sqrt{10}$. Therefore, $r = A/s = \sqrt{10}$.

Solution 3:

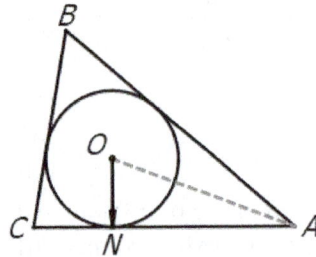

In $\triangle ABC$, let $AC = 14$, $AB = 13$, and $BC = 9$. Let N be the tangent point on AC. Then, $AN = (AC + AB - BC)/2 = (13 + 14 - 9)/2 = 9$.

The law of cosines gives $\cos A = (13^2 + 14^2 - 9^2)/(2 * 13 * 14) = 71/91$. Using the **Half Angle Formula**,

$$\tan(A/2) = \sqrt{\frac{1 - \cos A}{1 + \cos A}} = \sqrt{\frac{20}{162}} = \frac{\sqrt{10}}{9}$$

Therefore, the radius $r = AN * \tan(A/2) = \sqrt{10}$.

Problem 211

A triangle in the coordinate plane has vertices at $(-4, 0)$, $(6, 0)$ and $(0, 5)$. The line $y = 5/4x + c$, where c is a positive number, divides the triangle into a trapezoid and a smaller triangle whose areas, respectively, are in the ratio 5:4. What is the value of c? Express your answer as a common fraction. (Mathcounts 2017 State Sprint)

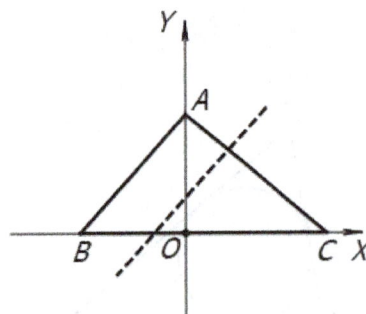

> **Tips:** *Calculate the ratio of the segments that are divided by the line using area ratios.*

Solution:

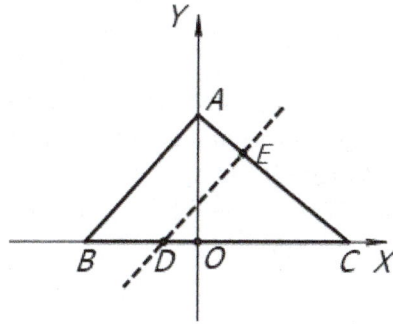

The coordinates of points A, B, and C are (-4, 0), (6, 0), and (0, 5), respectively. The slope of segment AB is equal to $5/4$. Therefore, $AB//DE$.

Because the line DE divides the triangle into two pieces with an area ratio of $5/4$, the area ratio of triangle CDE to triangle CAB is $4/9$. $AB//DE$ yields $\triangle CDE \sim \triangle CAB$. Thus, $CD/BC = 2/3$. Therefore, the coordinate of point D is (-2/3, 0).

Plugging the coordinates into the line equation $y = 5/4x + c$ yields $0 = -5/4 \times 2/3 + c$. Thus, $c = 5/6$.

Problem 212

Lines AB and DC are parallel, and transversals AC and BD intersect at a point X between the two lines so that $AX/CX = 5/7$. Points P and Q lie on segments AB and DC, respectively. The segment PQ intersects transversals BD and AC at points M and N, respectively, so that $PM = MN = NQ$. What is the ratio AP/BP? Express your answer as a common fraction. (Mathcounts 2017 State Sprint)

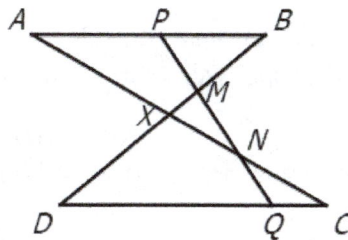

Tips: *Use the ratio information from parallel lines/similar triangles.*

Solution:

Let a and b be DQ and QC, respectively. Because $AB//CD$, $AB/CD = AX/XC = 5/7$; $AP/QC = PN/QN = 2$; $PB/DQ = PM/MQ = 1/2$.

Thus, $AP = 2b$, $PB = a/2$ and $AB = 5/7(a + b)$. So $2b + a/2 = 5/7(a + b)$. Simplifying the equation yields $a = 6b$. Therefore, $AP/BP = (2b)/(3b) = 2/3$.

Problem 213

In right triangle ABC with right angle at vertex C, a semicircle is constructed, as shown, with center P on leg AC, so that the semicircle is tangent to leg BC at C, tangent to the hypotenuse AB, and intersects leg AC at Q between A and C. The ratio of AQ to QC is $2:3$. If $BC = 12$, then what is the value of AC? Express your answer in simplest radical form. (Mathcounts 2017 National Sprint)

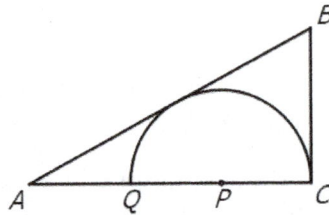

Tips: *Connect the center of the semicircle to the tangent point and use information from similar triangles.*

Solution 1:

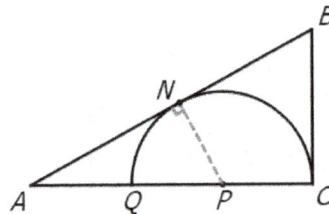

Let r be the radius of the circle. Because $AQ/QC = 2/3$, $AQ = \frac{4}{3}r$ and $AC = \frac{10}{3}r$. The **Power of a Point Theorem** yields $AN^2 = AQ * AC = \frac{40}{9}r^2$. Thus, $AN = \frac{2\sqrt{10}}{3}r$.

Connect point P to the tangent point N. $PN \perp AB$ and $\triangle ANP \sim \triangle ACB$. Therefore, $AN/AC = PN/BC$. $r = PN = AN/AC * BC = (\frac{2\sqrt{10}}{3}r)/(\frac{10}{3}r) * 12 = 12\sqrt{10}/5$.

Solution 2:

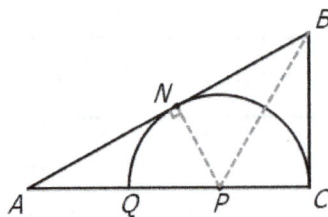

Because $AQ/QC = 2/3$ and $QP = PC$, $AP/PC = 7/3$.

Connect point P to the tangent point N. BP is the angular bisector of angle B. The **Angular Bisector Theorem** yields $AB/BC = AP/PC = 7/3$. Thus, $AB = (7/3) * BC = 28$ and $AN = AB - BN = 28 - 12 = 16$.

Let r be the radius of the circle, $AP = \frac{7}{3}r$. The **Pythagorean Theorem** in triangle ANP gives $(49/9 - 1)r^2 = 16^2$. Solving the equation yields $r = 12\sqrt{10}/5$.

Solution 3:

Similar to solution #2, because $AQ/QC = 2/3$ and $QP = PC$, $AP/PC = 7/3$.

Connect point P to the tangent point N. BP is the angular bisector of angle B. The **Angular Bisector Theorem** yields $AB/BC = AP/PC = 7/3$. Thus, $AB = (7/3) * BC = 28$ and $AN = AB - BN = 28 - 12 = 16$.

Let r be the radius of the circle, $AQ = \frac{4}{3}r$. And $AC = \frac{10}{3}r$. The **Power of a Point Rule** yields $(40/9) * r^2 = 16^2$. Solving the equation yields $r = 12\sqrt{10}/5$.

Problem 214

The diagonals of parallelogram $ABCD$ intersect at point E. Point F is the midpoint of the segment BE and point H is the midpoint of segment CE. What is the ratio of the area of quadrilateral $AFHD$ to the area of the parallelogram?

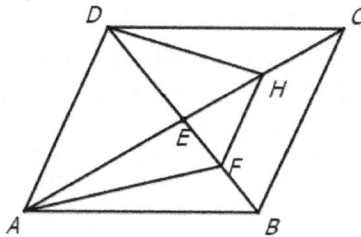

> **Tips:** *Use the ratio information from parallel lines/similar triangles.*

Solution:

The diagonals divided the parallelogram into 4 triangles with equal area. Let s be the area of one of the four small triangles. The area of the whole parallelogram is $4s$.

Because the points F and H are the midpoints of EC and EB, respectively, the area of triangle DEH, AEF and EFH is $s/2$, $s/2$ and $s/4$, respectively. The area of the trapezoid $AFHD$ is $9s/4$ and the ratio of the area of quadrilateral $AFHD$ to the area of the parallelogram is $9/16$.

Problem 215

In isosceles trapezoid $ABCD$, shown here, sides AB and DC are parallel, $AB = 10$ and $CD = 8$. Trapezoids $APQR$ and $BCQP$ are both similar to trapezoid $ABCD$. What is the

303

area of trapezoid $ABCD$? Express your answer in simplest radical form. (Mathcounts 2017 National Sprint)

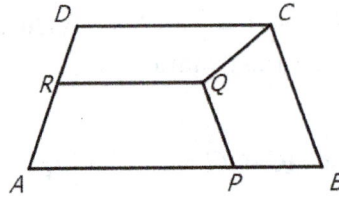

Tips: *1. Use the properties of similar shapes to find the length of the sides. 2. Create a right triangle to calculate the altitude of the trapezoid.*

Solution:

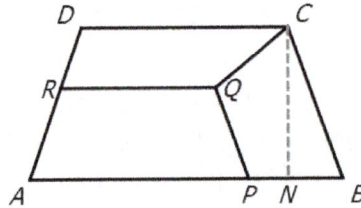

Because trapezoid $ABCD$ and $BCQP$ are similar, $QP/BC = CD/AB = 8/10$. Similarly, because trapezoid $ABCD$ and $APQR$ are also similar, $AP/AB = QP/BC = 8/10$. Thus, $AP = 8$ and $PB = 2$.

Let the length of BC be a. The ratio properties for the similar trapezoids $ABCD$ and $BCQP$ gives $a/10 = 2/a$. Solving the equation yields $a = 2\sqrt{5}$.

Construct a right triangle with one leg being the altitude CN. $NB = (10 - 8)/2 = 1$. The length of altitude $CN = \sqrt{19}$ and the area of the trapezoid is equal to $9\sqrt{19}$.

Problem 216

In the figure shown, two lines intersect at a right angle, and two semi circles are drawn so that each semicircle has its diameter on one line and is tangent to the other line. The larger semicircle has radius 1. The smaller semicircle intersects the larger semicircle, dividing the larger semicircular arc in the ratio 1 : 5. What is the radius of the smaller semicircle? Express your answer in simplest radical form. (Mathcounts 2017 National Sprint)

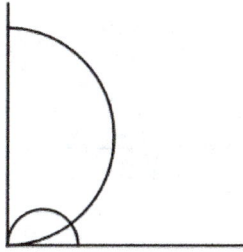

Tips: *1. Find the central angle of the chord. 2. Construct right triangle to calculate the radius of the small circle.*

Solution 1:

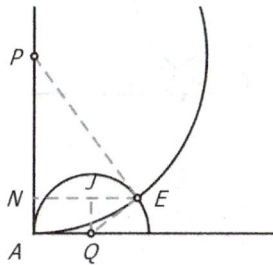

Let point N is on PA such that $EN \perp PA$ and point J be on NE such that $QJ \perp NE$.

Let r be the radius of the small semicircle. Because the small semicircle divides the big semicircle arc into $1 : 5$, $\angle APE = 30°$. Therefore, $NE = 1/2$ and $PN = \sqrt{3}/2$. Thus, $JQ = NA = 1 - \sqrt{3}/2$ and $JE = 1/2 - r$. In the right triangle JEQ, the **Pythagorean Theorem** yields $(1 - \sqrt{3}/2)^2 + (1/2 - r)^2 = r^2$.

Solving the equation yields $r = 2 - \sqrt{3}$.

Solution 2:

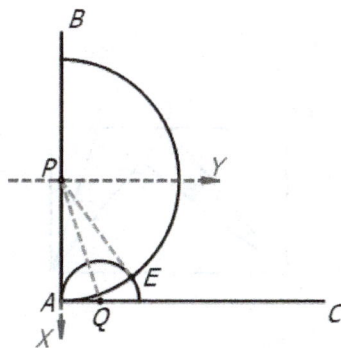

Define a coordinate system as shown in blue. Because the small semicircle divides the big semicircle arc into $1 : 5$, $\angle APE = 30°$. The coordinates of point E are $(\sqrt{3}/2, 1/2)$.

305

The equation for the small circle can be expressed as $(x - 1)^2 + (y - r)^2 = r^2$. The point E is on the circle, so $(\sqrt{3}/2 - 1)^2 + (1/2 - r)^2 = r^2$. Solving the equation gives

$$r = \sqrt{\frac{2 - \sqrt{3}}{2 + \sqrt{3}}} = 2 - \sqrt{3}$$

.

Solution 3:

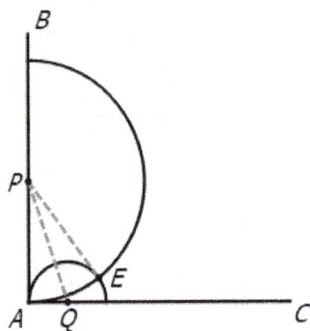

Because the small semicircle divides the big semicircle arc into $1 : 5$, $\angle APE = 30°$. PQ is the angular bisector of $\angle APE$, so $\angle APQ = 30°$.

Thus, the radius of the small semicircle is

$$AQ = PA \tan 15° = \tan 15° = \sqrt{\frac{1 - \cos 30°}{1 + \cos 30°}} = \sqrt{\frac{2 - \sqrt{3}}{2 + \sqrt{3}}} = 2 - \sqrt{3}$$

Problem 217

In rectangle $TUVW$, shown here, $WX = 4$ units, $XY = 2$ units, $YV = 1$ unit and $UV = 6$ units. What is the absolute difference between the areas of triangles TXZ and UYZ. (Mathcounts 2016 State Sprint)

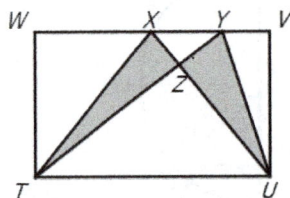

Tips: *1. Use the area method to compare the area of two triangles. 2. Apply the equal area property for the two side triangles between the two parallel lines.*

Solution 1:

Knowing that the area of the two side triangles created by crossing lines between two parallel lines is equal, the difference between the areas of triangles TXZ and UYZ is 0.

Solution 2:

Because $WV//TU$, $XZ/ZU = YZ/ZT$. Thus, $S_{\triangle TXZ}/S_{\triangle TZU} = XZ/ZU = YZ/ZT = S_{\triangle UYZ}/S_{\triangle TZU}$. Therefore, $S_{\triangle TXZ} = S_{\triangle UYZ}$ and their difference is 0.

Problem 218

What is the radius of a circle inscribed in a triangle with sides of length 5, 12 and 13 units? (Mathcounts 2015 State Sprint)

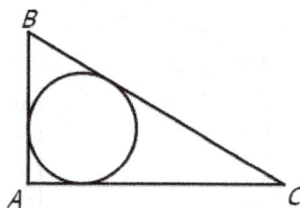

Tips: *1. Use the inscribed circle radius formula for right triangles. 2. Use the area of the triangle to calculate the radius of the inscribed circle.*

Solution 1:

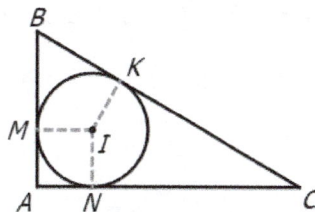

In a right triangle, the diameter of the inscribed circle is the difference between the sum of the two legs and the hypotenuse. Thus, $2r = 5 + 12 - 13 = 4$. The radius of the inscribed circle is 2.

Solution 2:

The area of the right triangle is $S = bc/2 = sr$, where $s = (a+b+c)/2 = (5+12+13)/2 = 15$. Therefore, $r = bc/(2s) = 5 \times 12/2/15 = 2$.

Problem 219

A rectangle of perimeter 22 cm is inscribed in a circle of area 16π cm^2. What is the area of the rectangle? Express your answer as a decimal to the nearest tenth. (Mathcounts 2015 State Sprint)

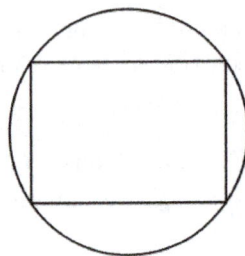

Tips: *The diagonal of the rectangle is the diameter of the circle.*

Solution:

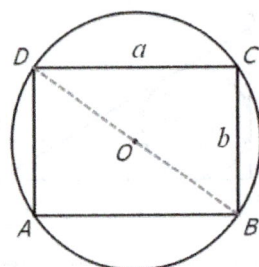

Because the area of the circle is 16π cm^2, the diameter of the circle BD is 8 cm.

Let $BC = b$ and $CD = a$. Because the perimeter is 22 cm, $a + b = 11$ (1). The **Pythagorean Theorem** in the right triangle BCD yields $a^2 + b^2 = 8^2$ (2). Squaring the first equation and subtracting equation 2 yields $ab = (121 - 64)/2 = 28.5$ cm^2, which is the area of the rectangle.

Problem 220

The hypotenuse of isosceles right triangle ABC is a side of square $ACDE$ as shown. If $AB = 4$ units, what is the length of segment BE? Express your answer in simplest radical form. (Mathcounts 2015 National Sprint)

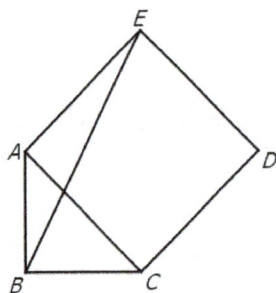

Tips: *1. Construct a right triangle with BE as one of the sides.*

Solution 1:

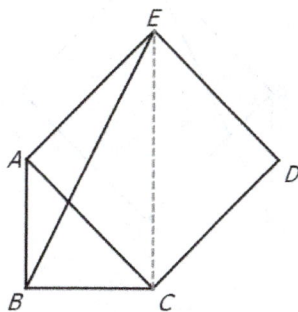

Connect the diagonal CE. Then, $\angle ECB = \angle ACB + \angle ECQ = 45° + 45° = 90°$. Because $AB = 4$, we have $AC = 4\sqrt{2}$, $BC = 4$ and $CD = 8$. Therefore, in the right triangle BCE, $BE = 4\sqrt{5}$.

Solution 2:

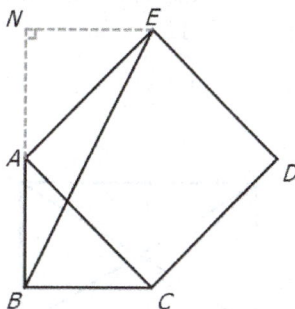

Let point N be on the extension of BA such that $EN \perp AN$. Triangle ANE is an isosceles right triangle and $\triangle ANE \cong \triangle ABC$. Thus, $NE = NA = AB = 4$. Therefore, in the right triangle BNE, $BE = 4\sqrt{5}$.

Solution 3:

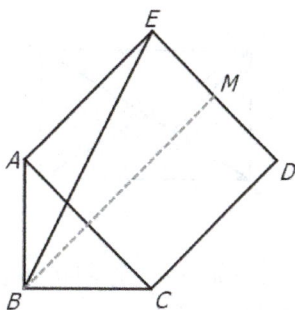

Let point M be the midpoint of DE and connect BM. Because the polygon $ABCDE$ is symmetric and M is the midpoint of DE, $BM \perp DE$. In the right triangle BEM, $EM = 2\sqrt{2}$ and $BM = 6\sqrt{2}$. Therefore, $BE = 4\sqrt{5}$.

Solution 4:

309

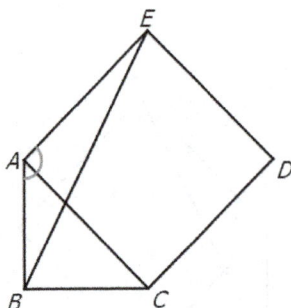

In triangle BAE, $AB = 4$, $AE = 4\sqrt{2}$, and $\angle BAE = 135°$. The Law of Cosines yields $BE^2 = 16 + 32 + 32$. Therefore, $BE = 4\sqrt{5}$.

Problem 221

The angular bisector of $\angle ADC$ intersects the diagonal AC at point E. Given $AD = 1$ unit and $\angle BDE = 15°$. Find the length of AE.

(A) $\sqrt{2}/2$ (B) $\sqrt{3}/3$ (C) 1   (D) $\sqrt{3} - 1$ (E) $\sqrt{2} - 1$

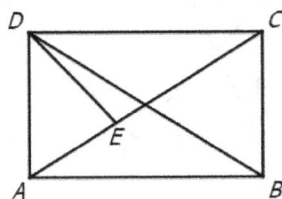

Tips: *1. Find the measure of angle CAB. 2. Use the angular bisector theorem.*

Solution 1:

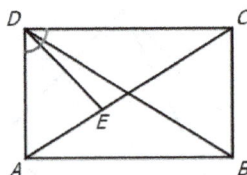

Because DE is the angular bisector of angle $\angle ADC$ and $\angle BDE = 15°$, $\angle ACD = 30°$. Thus, $DC = \sqrt{3}AD = \sqrt{3}$ and $AC = 2AD = 2$.

The **Angular Bisector Theorem** yields $AE/EC = AD/DC$.

$AE = AC * AD/(AD + DC) = 2/(1 + \sqrt{3}) = \sqrt{3} - 1$

The answer is (D).

Solution 2:

310

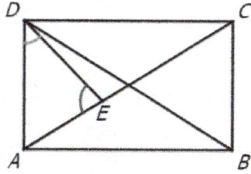

Because DE is the angular bisector of angle $\angle ADC$ and $\angle BDE = 15°$, we know that $\angle ACD = 30°$, $\angle ADE = 45°$, and $\angle AED = 75°$.

The law of sines in triangle ADE gives $AE/AD = \sin 45°/\sin 75° = \sin 45°/\sin 30° * \cos 45° + \cos 30° * \sin 45° = 1/(1/2 + \sqrt{3}/2) = \sqrt{3} - 1$

The answer is (D).

Solution 3:

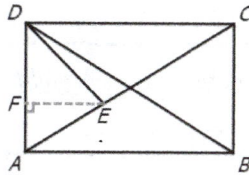

Because DE is the angular bisector of angle $\angle ADC$ and $\angle BDE = 15°$, we know that $\angle ACD = 30°$, $\angle ADE = 45°$ and $\angle DAE = 60°$.

Let point F be on AD such that $EF \perp AD$. Thus, triangle DEF is an isosceles right triangle. Let $x = AF$. Thus, $FE = DF = \sqrt{3}x$ and $AE = 2AF = 2x$. Given $AD = AF + DF = 1$, $x + \sqrt{3}x = 1$, and $x = (\sqrt{3} - 1)/2$. Thus, $AE = 2x = \sqrt{3} - 1$.

The answer is (D).

Solution 4:

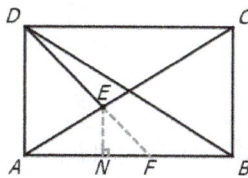

Extend DE to meet AB at point F and let point N be on AB such that $EN \perp AB$.

Because DE is the angular bisector of angle $\angle ADC$ and $\angle BDE = 15°$, $\angle ACD = \angle CAB = 30°$ and $\angle ADF = 45°$.

Let $x = NF$. Then, $EN = x$, $AN = \sqrt{3}x$ and $AE = 2x$. Given $AN + NF = AD = 1$, we have $x + \sqrt{3}x = 1$ and $x = (\sqrt{3} - 1)/2$. Thus, $AE = 2x = \sqrt{3} - 1$.

Answer is (D).

311

Problem 222

Point A is on the circle with BC as its diameter. D is on BC such that $AD \perp BC$. Point E is on BC and F is on the extension of CB, such that $\angle FAB = \angle CAE$. Find AE if $BC = 15$, $BF = 6$ and $BD = 3$.

(A) $4\sqrt{3}$ (B) $2\sqrt{13}$ (C) $2\sqrt{14}$ (D) $2\sqrt{15}$

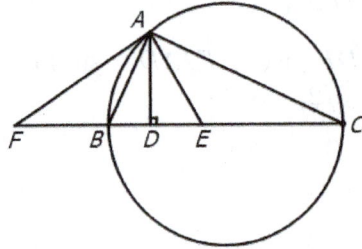

> **Tips:** *1. Segment AD is the altitude of two right triangles. 2. Use the property of the altitude of a right triangle.*

Solution:

BC is the diameter of the circle, so $\angle BAC = 90°$.

Because $\angle FAB = \angle CAE$, $\angle FAE = 90°$. Thus, $\triangle ABC$ and $\triangle FAE$ are right triangles and AD is the altitude of both right triangles.

Therefore, $AD^2 = FD * DE = BD * DC$. $DC = BC - BD = 15 - 3 = 12$ and $FD = BF + BD = 6 + 3 = 9$. Substituting into the equation gives us $AD = 6$ and $DE = 36/9 = 4$.

Therefore, in right triangle ADB, we have $AE = 2\sqrt{13}$.

The answer is (B).

Problem 223

The diagonal AC and BD of quadrilateral $ABCD$ intersects at point O. Given $BC = 3$, $AD = 4$, $AC = 5$ and $AB = 6$, find DO/OB if $\angle BAD + \angle ACB = 180°$.

(A) 10/9   (B) 8/7   (C) 6/5   (D) 4/3

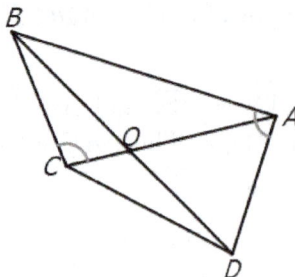

Tips: *1. The ratio of the two segments is equal to the ratio of the areas of the two triangles. 2. Use the trigonometry to calculate their areas.*

Solution 1:

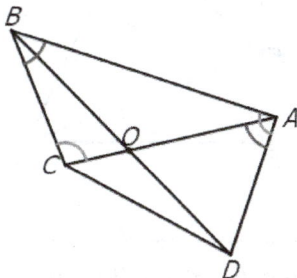

Because $\angle BAD + \angle ACB = 180°$, $\angle CAD = \angle BAD - \angle BAC = 180° - \angle ACB - \angle BAC = \angle CBA$.

Using the area method, $DO/OB = S_{\triangle ACD}/S_{\triangle ABC}$

$= ((1/2)*AC*BD\sin\angle CAD)/((1/2)*BC*AB\sin\angle CBA)$

$= AC*AD/AB*BC = 45/36 = 10/9$.

The answer is (A).

Solution 2:

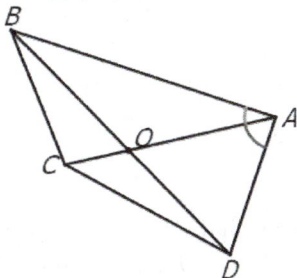

Because $\angle BAD + \angle ACB = 180°$, $\angle CAD = \angle BAD - \angle BAC = 180° - \angle ACB - \angle BAC = \angle CBA$. In triangle ABC, the law of sines yields $\sin\angle ABC/\sin\angle BAC = AC/BC = 5/3 = \sin\angle ABC/\sin\angle CAD$.

Using the area method, $DO/OB = S_{\triangle OAD}/S_{\triangle OAB} = AD*\sin\angle CAD/(AB*\sin\angle BAC) = 4/6 \times 5/3 = 10/9$.

The answer is (A).

Problem 224

In a right trapezoid $ABCD$, $AD//BC$, $\angle A = 90°$, $AB = BC = 70$. E is on side AB such that $\angle DCE = 45°$. Find DE.

(A) 56   (B) 58   (C) 60   (D) 62

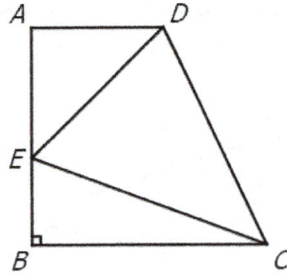

> **Tips:** *1. Construct a square based on the information. 2. Use 45 degree angle to construct a pair of congruent triangles.*

Solution 1:

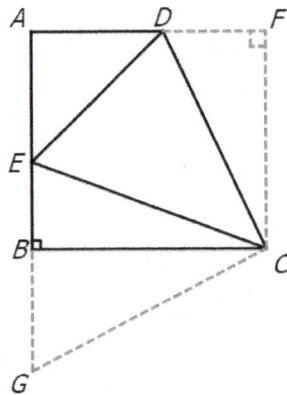

Let point F be on line AD such that $CF \perp AD$. $ABCF$ is a square with side-length of 70. Extend AB to G such that $BG = DF$. We have $\triangle CDF \cong \triangle CGB$. Thus, $CD = CG$, $DF = BG$ and $\angle FCD = \angle BCG$. Because $\angle DCE = 45°$, $\angle FCB = 90°$ and $\angle FCD = \angle BCG, \angle DCE = \angle GCE$.

Using the **SAS congruence rule**, we have $\triangle DCE \cong \triangle GCE$. Thus, $DE = EG = EB + DF$.

Let $x = DF$, then $AD = 70 - x$ and $DE = 28 + x$. In the right triangle ADE, the **Pythagorean Theorem** yields $42^2 + (70 - x)^2 = (28 + x)^2$. Solving the quadratic equation yields $x = 30$, so $DE = 28 + 30 = 58$.

The answer is (B).

Solution 2:

314

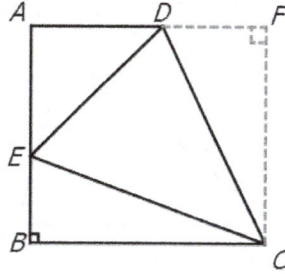

Let F be a point on line AD such that $CF \perp AD$. $ABCF$ is a square with side-length of 70.

Because $\angle DCE = 45°$, $\angle FCB = 90°$, $\angle ECB + \angle DCF = 45°$. It is straightforward to find that $\tan \angle ECB = 28/70 = 2/5$.

Thus, $\tan \angle DCF = \tan(45° - \angle ECB) = (1 - 2/5)/(1 + 1 \times 2/5) = 3/7$.

So $DF = CF \tan \angle DCF = 70 \times 3/7 = 30$ and $AD = 70 - 30 = 40$. Therefore, in the right triangle ADE, $DE = \sqrt{42^2 + 38^2} = 58$.

The answer is (B).

Solution 3:

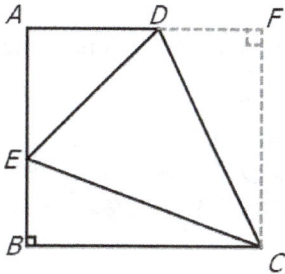

Let point F be on line AD such that $CF \perp AD$. $ABCF$ is a square with side-length of 70.

Let $x = DF$, then $AD = 70 - x$ and $CD^2 = 70^2 + x^2$ and $CE^2 = 70^2 + 28^2$.

The law of cosines in triangle CDE yields $DE^2 = CE^2 + CD^2 - 2CE * CD * \cos 45° = 70^2 + x^2 + 70^2 + 28^2 - \sqrt{2(70^2 + x^2)(70^2 + 28^2)}$. In the right triangle ADE, we have $DE^2 = AE^2 + AD^2 = 42^2 + (70 - x)^2$.

Therefore $70^2 + x^2 + 70^2 + 28^2 - \sqrt{2(70^2 + x^2)(70^2 + 28^2)} = 42^2 + (70 - x)^2$. Carefully simplifying the equation and solving the equation yields $x = 30$.

Therefore, $DE = \sqrt{42^2 + 38^2} = 58$.

The answer is (B).

Problem 225

In a right trapezoid $ABCD$, $AD//BC$, $AB = 3$, $BC = 4$, $CD = 2$ and $AD = 1$. Find the area of the trapezoid $ABCD$.

315

(A) $10\sqrt{2}/3$ (B) $10\sqrt{3}/3$ (C) $3\sqrt{2}$ (D) $3\sqrt{3}$

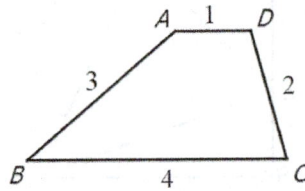

> **Tips:** *1. Construct a triangle parallel to one of the sides and calculate the altitude of that triangle.*

Solution 1:

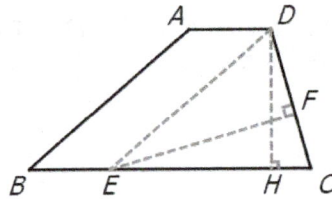

Draw $DE//AB$ where E is a point on BC. Then, $AB \perp ED$ is a parallelogram and $BE = AD = 1$ and $DE = AB = 3$.

In triangle CDE, $DE = CE = 3$ and $CD = 2$. Thus, the altitude EF of side CD is equal to $2\sqrt{2}$. The altitude on side BC is $DH = EF * CD/CE = 2/3 \times 2\sqrt{2} = 4\sqrt{2}/3$. Therefore, the area of the trapezoid is equal to $(1+4)/2 \times 2\sqrt{2}/3 = 10\sqrt{2}/3$.

The answer is (A).

Solution 2:

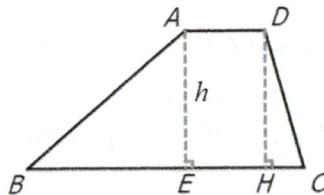

Let points E and H be on side BC such that $AE \perp BC$ and $DH \perp BC$. Then, $ADHE$ is a rectangle and we have the two right triangles ABE and CDH.

Let $h = AE = DH$. Then, $BE = \sqrt{3^2 - h^2}$ and $CH = \sqrt{2^2 - h^2}$. We also have $BE + CH = BC - EH = 3$. Therefore, $\sqrt{3^2 - h^2} + \sqrt{2^2 - h^2} = 3$ (1).

Solving equation (1) yields $h = 4\sqrt{2}/3$.

Therefore, the area of the trapezoid is equal to $(1+4)/2 \times 2\sqrt{2}/3 = 10\sqrt{2}/3$.

The answer is (A).